Approximation on a rectangular grid

Monographs and textbooks on mechanics of solids and fluids
editor-in-chief: G. Æ. Oravas

Mechanics: Analysis
editor: V. J. Mizel

1. M. A. KRASNOSELSKII *et al.*
   Integral operators in spaces of summable functions

2. V. V. IVANOV
   The theory of approximate methods and their application to the numerical solution of singular integral equations

3. A. KUFNER *et al.*
   Function spaces

4. S. G. MIKHLIN
   Approximation on a rectangular grid

Approximation on a rectangular grid

with application to finite element methods and other problems

S. G. Mikhlin

Mathematics Department
Leningrad State University

Translated and edited by

R. S. Anderssen

Pure Mathematics, School of General Studies, Australian National University
and
Division of Mathematics and Statistics, Commonwealth Scientific and Industrial Research Organization

and

T. O. Shaposhnikova,
(Leningrad)

Sijthoff & Noordhoff 1979
Alphen aan den Rijn The Netherlands
Germantown, Maryland USA

ISBN 90 286 0008 6

Original title:
Approksimatcija na kubichcskoj setke

Printed in The Netherlands

PREFACE

The present monograph has points in common with two branches of analysis. One of them is the variational-difference method (the finite element method), the other is the constructive theory of functions. The starting point is the construction of special classes of coordinate functions for the variational-difference method. It is based on elementary transformations of the independent variables of given "primitive" functions. After the construction of the coordinate functions, the next step is to approximate functions of a given class by linear combinations of the coordinate functions, and to derive in some appropriate norm an estimate of the error. Clearly, this is a problem closely connected with the constructive theory of functions.

The monograph contains 11 chapters. Chapter I discusses Courant's basic idea which is central to the construction of variational-difference methods. One of Courant's examples, from which the notion of a primitive function follows naturally, is examined in some detail. The general definition of a primitive function and the method of construction for the corresponding coordinate functions are given and discussed.

Chapters II-VI are more closely connected with the constructive theory of functions. The completeness of the coordinate systems defined in Chapter I are studied, as well as the order of approximation obtained through the use of linear combinations of these functions. Their completeness in Sobolev spaces are examined in Chapter II, while related orders of approximation are derived in Chapter III. Approximations based on the use of the "primitive functions with wide support" are studied in Chapter IV. Attention is chiefly restricted to the one-dimensional situation. The accuracy of the approximations obtained when used to solve ordinary and partially degenerate differential equations is the topic of Chapters V and VI. The metric examined is usually the appropriate energy norm.

An examination of the variational-difference method is given in Chapters VII-IX. The major result of Chapter VII is the conclusion that the largest approximate eigenvalue of a positive definite operator with a discrete spectrum, calculated using a variational-difference method, has the same order of growth as the corresponding exact eigenvalue. Additional theorems about approximate eigenvalues are included.

Several methods for the construction of variational-difference (algebraic) systems for various types of differential equations and boundary conditions are developed in Chapter VIII. In particular, certain variational-difference schemes which "contain a boundary layer" are constructed. Compared with the usual systems, these schemes yield algebraic systems with fewer unknowns. The stability of variational-difference processes as well as the condition number of the corresponding matrices are investigated in Chapter IX. Methods for solving the algebraic systems, which variational-difference methods generate, are not discussed in this book, since excellent texts which treat this topic are widely available. A numerical example is given at the end of Chapter IX.

Some special quadrature and cubature formulas are presented in Chapter X. Every approximation of a function obviously yields formulas for the approximate evaluation of integrals. The approximations of the first chapters yield, in one-dimension, the well known Euler-Maclaurin sum formula. There are corresponding analogues in multidimensions. Replacing derivatives by differences, cubature formulas, analogous to the well known Gregory quadrature formula, are obtained in the usual manner.

In Chapter XI the variational-difference method is used for the approximate solution of second kind Fredholm integral equations. The necessity to reduce the problem to an algebraic system is eliminated. It is sufficient to construct a variational-difference approximation for the numerator and denominator of the Fredholm resolvent kernel using the corresponding approximation of the kernel itself. Considerable attention has been paid to estimating the absolute and probable error.

The monograph is based principally on results of the author. A small number of results of other authors are also used.

It is assumed that the reader is familiar with the elements of the functional analysis (including imbedding theorems for Sobolev spaces) as well as with variational and finite-difference methods.

S. Mikhlin,
Leningrad, 1978.

THE TRANSLATORS' PREFACE

The translators would like to take this opportunity to record their thanks to people and institutions who have helped in some way or other with the preparation of this book. In particular, we wish to record our thanks to Professor S.G. Mikhlin for his willingness to advise and assist with the translation when his help was needed, to Mrs Stephanie Venema, Sijthoff and Noordhoff, The Netherlands, for her assistance with the publication, to Professor J.R.M. Radok, The Horace Lamb Institute of Oceanography, for his inspiration, and to Mrs Barbara Geary, Pure Mathematics, IAS, Australian National University, for her superior typing. In addition, thanks also go to Graeme Chandler, Mark Lukas, Andreas Griewank, Kris Jittorntrom and John Paine (PhD students in numerical analysis at the Australian National University) and Clementine Krayshek, RSES, Australian National University (who assisted with the preparation of the Figures), as well as Neil Trudinger, Pure Mathematics, School of General Studies.

The book itself is not an elementary introduction. Graduate level experience with variational methods, functional analysis, Sobolev spaces and pde is assumed.

Robert S. Anderssen,
Canberra, 1979.

T.O. Shaposhnikova,
Leningrad, 1979.

TABLE OF CONTENTS

CHAPTER I

THE PRIMITIVE FUNCTIONS

§1. The Variational-Difference Method

The original formulation of the variational-difference method is due to Courant [1]. We reformulate it below using the more recent terminology of applied functional analysis.

Consider a *positive definite operator* A which maps a separable Hilbert space into itself. It is assumed that A is linear. The *aim* is to construct for a given f, the generalized solution u_0 of the corresponding operator equation

$$\mathsf{A}u = f \qquad (f \in H) . \tag{1}$$

From Mikhlin [1], § , it is known that this generalized solution yields the minimum of the *energy functional*

$$F(u) = \|u\|_{\mathsf{A}}^2 - 2(u, f) , \tag{2}$$

in the *energy space* H_{A} of the operator A, where $\|u\|_{\mathsf{A}}^2 = (\mathsf{A}u, u)$ denotes the *energy norm;* that is, the norm in the space H_{A}. The *Ritz method* can be used to generate approximations to the solution of this minimization problem: for a given sequence of finite-dimensional subspaces $H_{\mathsf{A}}^{(n)} \subset H_{\mathsf{A}}$, $n = 1, 2, \ldots$, such that $H_{\mathsf{A}}^{(n)} \subset H_{\mathsf{A}}^{(n+1)}$, $n \geq 1$, construct the approximate solution u_n (the approximate Ritz solution) as the element in H_{A} at which the energy functional (2) attains its minimum on $H_{\mathsf{A}}^{(n)}$. If a basis $\varphi_1, \varphi_2, \ldots, \varphi_{k_n}$, for $H_{\mathsf{A}}^{(n)}$ is chosen, with k_n equal to the dimension of

$H_A^{(n)}$, then u_n takes the form

$$u_n = \sum_{k=1}^{k_n} a_k \varphi_k \ , \quad a_k = \text{scalar}. \tag{3}$$

The condition that u_n be the element in H_A at which the energy functional (2) attains its minimum on $H_A^{(n)}$ now yields the algebraic system (the Ritz-system)

$$\sum_{k=1}^{k_n} a_k [\varphi_k, \varphi_j]_A = (f, \varphi_j) \ , \quad j = 1, 2, \ldots, k_n \ , \quad [u, v]_A = (Au, v) \ . \tag{4}$$

Normally, in the Ritz method, it is not the subspaces $H_A^{(n)}$ which are given, but a countable spanning sequence of elements $\varphi_k \in H_A$, $k = 1, 2, \ldots$, and an increasing sequence of non-negative integers k_n, $n = 1, 2, \ldots$. The subspaces $H_A^{(n)}$ are thereby defined as the subspaces of H_A with bases $\{\varphi_1, \varphi_2, \ldots, \varphi_{k_n}\}$, $n = 1, 2, \ldots$. Clearly, the condition that $H_A^{(n)} \subset H_A^{(n+1)}$, $n \geq 1$, is automatically satisfied, as the $\{\varphi_k\}$ are linearly independent. The elements φ_k are called *coordinate functions*.

In his fundamental paper, Courant [1] pointed out that it is not necessary to demand that $H_A^{(n)} \subset H_A^{(n+1)}$, $n \geq 1$. By removing this condition, the range of choice for $H_A^{(n)}$ is extended. Through the use of specific examples, Courant showed that the subspaces $H_A^{(n)}$ could be chosen so that the matrix of the Ritz-system (4) coincides with the matrix of a specific finite-difference scheme. However, the right hand side values of the Ritz-system (4) will differ slightly from the corresponding values of the finite-difference scheme.

In order to construct the Ritz-system (4) in this new context, it is necessary to choose a basis for each of the subspaces $H_A^{(n)}$, separately. For convenience, two identifiers will be used to denote elements of such bases: a superscript n to identify the sub-space $H_A^{(n)}$, and a subscript k to identity the individual elements. If, as before, the dimension of the subspace $H_A^{(n)}$ is taken to be k_n, then the new representation for the basis generating $H_A^{(n)}$ becomes

$$\varphi_1^{(n)}, \varphi_2^{(n)}, \ldots, \varphi_{k_n}^{(n)} . \tag{5}$$

These elements are also called *coordinate functions*. Since the coordinate functions belong to one of the subspaces $H_A^{(n)}$, they also belong to the energy space H_A of the operator A. We also note that, for n fixed, the functions (5) are linearly independent as they form a basis for $H_A^{(n)}$.

We say that the sequence of subspaces $\{H_A^{(n)}\}$ is *complete* in H_A, if, for any element $u \in H_A$,

$$\lim_{n\to\infty} \inf_{v_n \in H_A^{(n)}} \| u-v_n \|_A = 0 . \tag{6}$$

This condition is equivalent to the following: for given $u \in H_A$ and $\varepsilon > 0$, there exists $N = N(u, \varepsilon)$ such that, for any $n > N$, there is an element $v_n \in H_A^{(n)}$ satisfying

$$\| u-v_n \|_A < \varepsilon . \tag{7}$$

We also say that such a system of coordinate functions (or, more briefly, a *coordinate system*)

$$\{\varphi_k^{(n)}\} , \quad n = 1, 2, \ldots , \quad k = 1, 2, \ldots, k_n , \tag{8}$$

is *complete* in H_A.

If the sequence $\{H_A^{(n)}\}$ is complete in H_A, then the approximate solutions (3), constructed by means of the Ritz-system (4), converge to the exact solution u_0. Indeed (see Mikhlin [3]), $F(u_n) = \| u_n-u_0 \|_A^2 - \| u_0 \|_A^2$ where u_n minimizes $\| \theta_n-u_0 \|_A^2$ for $\theta_n \in H_A^{(n)}$. For given $\varepsilon > 0$, we find $N(u_0, \varepsilon)$ and, for any $n > N$, we find $v_n \in H_A^{(n)}$ such that inequality (7) holds with $u = u_0$. Then

$$\| u_0-u_n \|_A \le \| u_0-v_n \|_A < \varepsilon ,$$

which completes the proof.

§2. An Example

We consider in detail one of Courant's examples. Let Ω be a finite domain in the plane with boundary $\partial\Omega = \Gamma$, and examine the use of the Ritz method for the construction of approximations to the solution of the following operator equation

$$-\Delta u \equiv -\left(\frac{\partial^2 u}{\partial x^2} + \frac{\partial^2 u}{\partial y^2}\right) = f(x, y) \ , \quad u|_\Gamma = 0 \ . \tag{1}$$

For a chosen integer n, set $h = 1/2n$ and construct a rectangular grid, with mesh size h, such that one of the mesh points coincides with the origin of the coordinate system and the sides of the rectangular grid are parallel to the coordinate axes. Consequently, the coordinates of the grid (mesh) points are (jh, kh), where j and k are integers. Each individual square of this rectangular grid is divided into two triangles by drawing in the diagonals, which are parallel to the line $y = x$. As well as the grid of "small" squares with side h, we also consider the grid of "larger" squares with side $2h$. Let $\hat{\Omega}^h$ denote the union of all the larger squares which can be placed with overlapping inside Ω, as illustrated in Figure 1. As $H^{(n)}$, Courant considered the subspace of functions which are continuous with respect to x and y, are linear in each triangle which belongs to the domain $\hat{\Omega}^h$, and vanishes outside Ω^h.

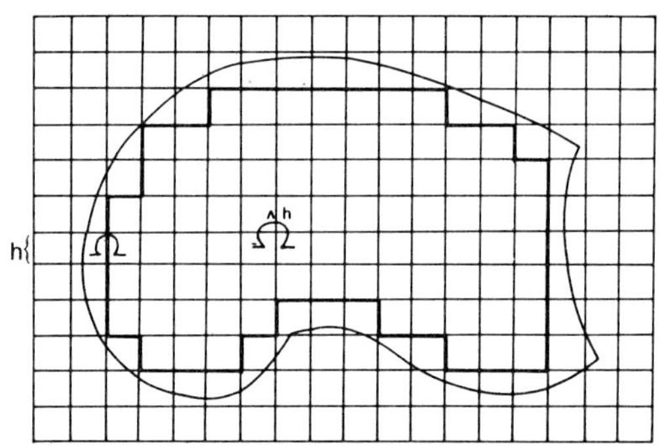

Fig. 1

It is not difficult to construct a basis for this subspace $H_A^{(n)}$. For each small (or larger) grid square, we define the "lower corner" point as that grid point for which the corresponding sides of the square have the same direction as the coordinate axes. We denote by $\hat{\mathcal{J}}^h$ the set of lower corner pairs $\{(j, k)\}$ such that the $\{(jh, kh)\}$ define the coordinates of the lower corner points of the larger squares in $\hat{\Omega}^h$. We now examine the gird points (vertices) of the small squares lying inside $\hat{\Omega}^h$. Let $\big((j+1)h, (k+1)h\big)$ denote the grid point lying at the center of the larger square with lower corner point (jh, kh). We construct the function $\big($in $H_A^{(n)}\big)$ which equals unity at this grid point and equals zero at all other grid points, and denote it by $\varphi_{jk}^{(n)}(x, y)$. Because of the linearity assumption, this function equals

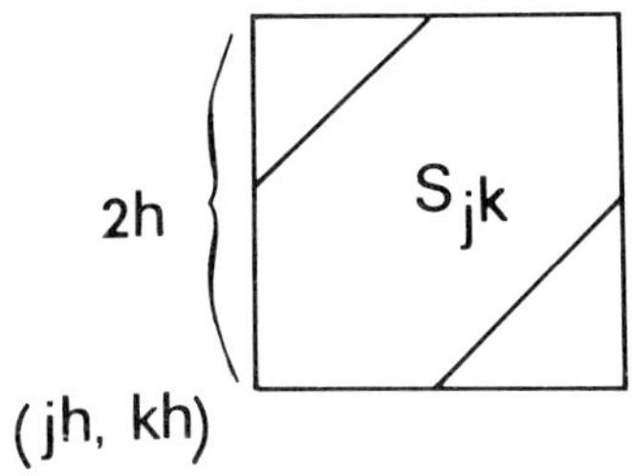

Fig. 2

zero outside the hexagon S_{jk} of Figure 2 and must be piecewise linear inside this hexagon. The surface $z = \varphi^{(n)}_{jk}(x, y)$ coincides with the plane $z = 0$ outside S_{jk}, and is the hexagonal pyramid with base S_{jk} and vertex $\big((j+1)h,\ (k+1)h,\ 1\big)$, inside S_{jk}. We now prove that the functions $\varphi^{(n)}_{jk}(x, y)$, $(j, k) \in \hat{J}^h$, form a basis for Courant's space $H^{(n)}_{\mathrm{A}}$. That is, we prove that, if $v \in H^{(n)}_{\mathrm{A}}$, then

$$v(x, y) = \sum_{(j,k)\in\hat{J}^h} v\big((j+1)h,\ (k+1)h\big)\varphi^{(n)}_{jk}(x, y) . \tag{2}$$

In fact, the right hand side of (2) is clearly a linear function on each of the triangles of Figure 1 and is continuous on $\hat{\Omega}^h$. Thus, in order to prove the identity (2), it only remains to verify that both sides are equal at grid points. For $x = (j_0+1)h$, $y = (k_0+1)h$, all $\varphi^{(n)}_{jk}(x, y)$ functions on the right hand side of (2) vanish except for $\varphi^{(n)}_{j_0,k_0}(x, y)$, and hence, the right hand side of (2) equals $v\big((j_0+1)h,\ (k_0+1)h\big)$, which establishes the required result.

It should be noted that all the functions $\varphi^{(n)}_{jk}(x, y)$ can be obtained from a given primitive function by a simple change of variables. For the square $0 \le x, y \le 2$, devide it into a hexagon S and two triangles by lines $y = x \pm 1$, as in Figure 2. We now define the following "primitive function": Its surface $z = \omega(x, y)$ coincides with the plane $z = 0$ outside S, and with the hexagonal pyramid with base S and vertex $(1, 1, 1)$ inside S. Clearly,

$$\varphi_{jk}^{(n)}(x, y) = \omega\left(\frac{x}{h} - j, \frac{y}{h} - k\right) . \tag{3}$$

The functions $\omega(x, y)$ are usually called *pyramid or hat* functions.

We seek an approximation to the solution of (1) in the form

$$u_n(x, y) = \sum_{(j,k)\in\hat{J}^h} a_{jk}\varphi_{jk}^{(n)}(x, y) . \tag{4}$$

Applying the Ritz method to the energy variational formulation for (1) (see Mikhlin [3]), we obtain the following system of equations

$$\sum_{(j,k)\in\hat{J}^h} \left[\varphi_{jk}^{(n)}, \varphi_{j_0k_0}^{(n)}\right]_A a_{jk} = \left(f, \varphi_{j_0k_0}^{(n)}\right) , \quad (j_0, k_0) \in \hat{J}^h , \tag{5}$$

for the unknowns a_{jk} .

The round and square brackets in (5) denote the scalar products in $L_2(\Omega)$ and H_A , respectively, where the space H_A is the energy space of the operator defined by (1):

$$(\varphi, \psi) = \int_\Omega \varphi(x, y)\psi(x, y)dxdy ,$$

$$[\varphi, \psi]_A = \int_\Omega \left(\frac{\partial\varphi}{\partial x}\frac{\partial\psi}{\partial x} + \frac{\partial\varphi}{\partial y}\frac{\partial\psi}{\partial y}\right)dxdy .$$

We examine the matrix of the algebraic system (5). It is clear that, if the larger squares with centers $\left((j_0+1)h, (k_0+1)h\right)$ and $\left((j+1)h, (k+1)h\right)$ do not intersect, then

$$\left[\varphi_{jk}^{(n)}, \varphi_{j_0k_0}^{(n)}\right]_A = 0 .$$

Consequently, every equation in (5) involves no more than nine unknowns; that is, the matrix of (5) consists of no more than nine nontrivial diagonals. One of them (the main diagonal) corresponds to the situation when $j_0 = j$ and $k_0 = k$; that is, when the larger squares coincide. The relationship between the abovementioned larger squares for each of the other eight diagonals is shown in Figures 3-6.

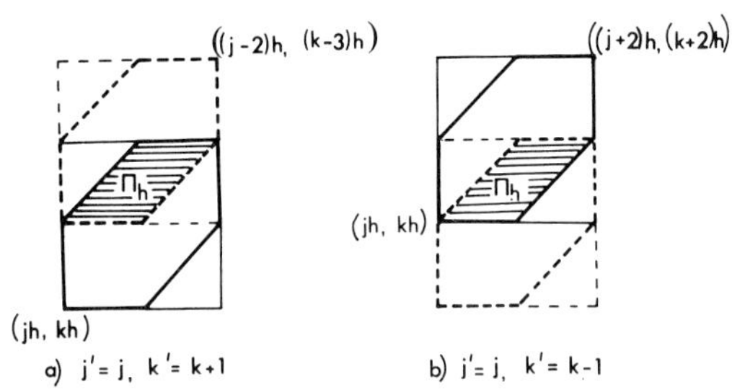

Fig. 3

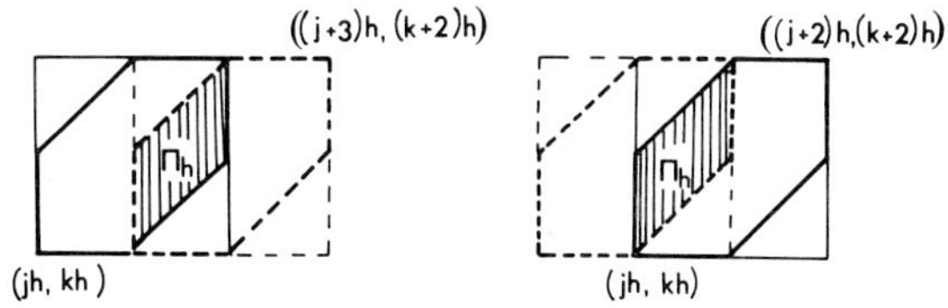

a) $j' = j+1, \quad k' = j$ b) $j' = j-1, \quad k' = k$

Fig . 4

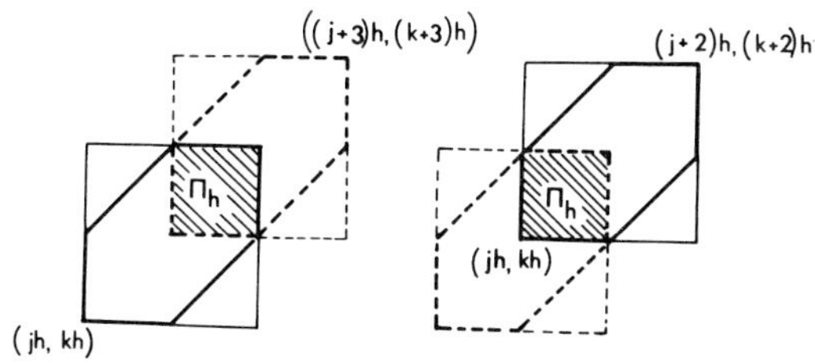

a) $j' = j+1, \quad k' = k+1$ b) $j' = j-1, \quad k' = k-1$

Fig. 5

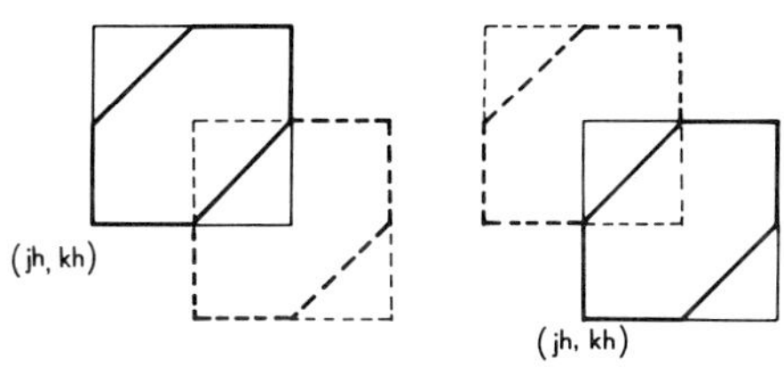

a) $j' = j+1, \quad k' = k-1$ b) $j' = j-1, \quad k = k+1$

Fig . 6

Let $j_0 = j$ and $k_0 = k$. Then

$$\left[\varphi_{jk}^{(n)}, \varphi_{jk}^{(n)}\right]_A = \int_{hj}^{(j+2)h} \int_{kh}^{(k+2)h} \left\{\left[\frac{\partial}{\partial x}\,\omega\left(\frac{x}{h} - j, \frac{y}{h} - k\right)\right]^2 + \left[\frac{\partial}{\partial y}\,\omega\left(\frac{x}{h} - j, \frac{y}{h} - k\right)\right]^2\right\} dxdy$$

$$= \int_0^2 \int_0^2 \left[\omega_x^2(x, y) + \omega_y^2(x, y)\right] dxdy = 4 \int_{T_1 \cup T_2} \left[\omega_x^2(x, y) + \omega_y^2(x, y)\right] dxdy ,$$

where the triangles T_1 and T_2 are illustrated in Figure 7. The function $\omega(x, y) = \varphi_{jh}^{(n)}(x, y)$ is linear on each of these triangles. When the origin is placed at the center $\left((j+1)h, (k+1)h\right)$ of the larger square, the corresponding planes are uniquely determined by the point sets $\{(0, 0, 1), (1, 1, 0), (-1, 1, 0)\}$ and $\{(0, 0, 1), (0, 1, 0), (-1, 0, 0)\}$, respectively. The equations defining these planes are therefore $z = 1 - y$ and $z = x - y + 1$. Hence,

$$\left[\varphi_{jk}^{(n)}, \varphi_{jk}^{(n)}\right]_A = 4\left\{\int_0^1 dx \int_x^1 dy + 2\int_{-\frac{1}{2}}^0 dx \int_{-x}^{1+x} dy\right\} = 4 . \tag{6}$$

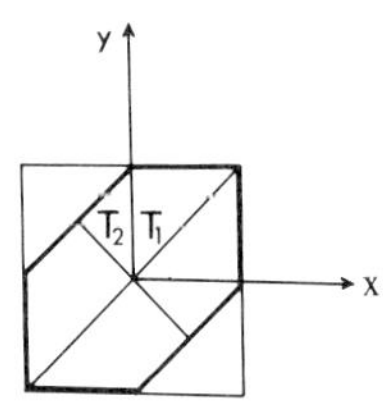

Fig. 7

For the situations illustrated in Figures 5 and 6, $\left[\varphi_{j_0k_0}^{(n)}, \varphi_{jk}^{(n)}\right] = 0$. It is obvious for Figure 6, since the supports of the pyramid functions in the integrand of $\left[\varphi_{jk}^{(n)}, \varphi_{j_0k_0}^{(n)}\right]_A$ intersect only along boundary lines. For Figure 52, a simple calculation verifies the anti-symmetry of the slopes of the planes entering the integrand of $[\varphi_{jk}, \varphi_{j_0k_0}]_A$. Consequently, it follows that the matrix of (5) is a banded matrix with a maximum of 5 non-trivial diagonals.

We now calculate the elements of the matrix of (5) for the situations illustrated in Figures 3, 4. Appealing to symmetry, it is clear that all these matrix elements are equal. It is therefore sufficient to make the calculations for the situation illustrated in Figure 3a. The supports of the functions

$$\varphi_{jk}^{(n)}(x, y) = \omega\left(\frac{x}{h} - j, \frac{y}{h} - k\right) \quad \text{and} \quad \varphi_{j_0k_0}^{(n)}(x, y) = \omega\left(\frac{x}{h} - j_0, \frac{y}{h} - k_0\right)$$

have the common intersection Π_h (the shaded parallelogram of Figures 3-4), over which it suffices to integrate $\left[\varphi_{jk}^{(n)}, \varphi_{j_0k_0}^{(k)}\right]$ to yield

$$\left[\varphi_{jk}^{(n)}, \varphi_{j_0k_0}^{(n)}\right]_A = \int_{\Pi_h}\int \left[\frac{\partial}{\partial x}\omega\left(\frac{x}{h} - j, \frac{y}{h} - k\right)\frac{\partial}{\partial x}\omega\left(\frac{x}{h} - j_0, \frac{y}{h} - k_0\right) + \frac{\partial}{\partial y}\omega\left(\frac{x}{h} - j, \frac{y}{h} - k\right)\frac{\partial}{\partial y}\omega\left(\frac{x}{h} - j_0, \frac{y}{h} - k_0\right)\right] dxdy .$$

Putting $\frac{x}{h} - j = \xi$ and $\frac{y}{h} - k = \eta$ and then changing ξ and η to x and y, we obtain

$$\left[\varphi_{jk}^{(n)}, \varphi_{j_0k_0}^{(n)}\right]_A = \iint_{\Pi} \left[\omega_x(x, y)\omega_x(x, y-1)+\omega_y(x, y)\omega_y(x, y-1)\right]dxdy , \tag{7}$$

where Π is the shaded parallelogram of Figure 8. The diagonal, parallel to y-axes, divides Π into two triangles T_3 and T_4. We evaluate (7) on each of these triangles separately.

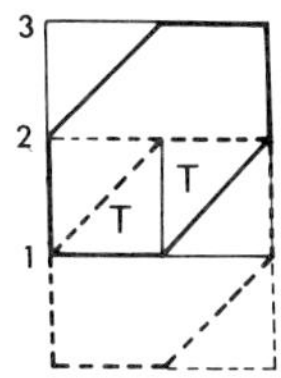

Fig. 8

On the triangle T_3, the function $\omega(x, y)$ is the plane which passes through the points $(0, 1, 0)$, $(1, 1, 1)$, $(1, 2, 0)$, so its equation is $z = x - y + 1$. Consequently, $\omega_x(x, y) = 1$ and $\omega_y(x, y) = -1$. Since the function $\omega(x, y-1)$ on the same triangle T_3 will be the plane which passes through the points $(0, 1, 0)$, $(1, 1, 0)$, $(1, 2, 1)$, it follows that $\omega(x, y-1) = y - 1$, and hence that $\omega_x(x, y-1) = 0$ and $\omega_y(x, y-1) = 1$. Consequently, integrating (7) on T_3 yields "$-\frac{1}{2}$". In the same way, (7) can be evaluated on T_4. In this way, we find that the value of (7) is "-1". Dividing each equation of (5) by "4", we obtain the standard five-point scheme for the non-homogeneous Laplace equation (that is, the Poisson equation), but with slightly modified right hand side values.

§3. The Basic Properties of Variational-Difference Matrices

The approximate solution of boundary value problems for both ordinary and partial differential equations by finite difference methods has certain advantages which motivate their extensive use in numerical applications, despite their known short comings: the necessity to solve high order algebraic systems even when the resulting approximations are not of high accuracy; the fast growth of the condition number of the grid matrix (that is, the general matrix corresponding to the matrix of (5)) as the order of the matrix increases; in some situations, the loss of certain useful properties of the differential equation, when it is replaced by a difference scheme.

In particular, there are situations where the difference scheme constructed for a given elliptic differential equation is not elliptic (see Frank [1]).

The variational-difference method is apparently free from the mentioned shortcomings. We consider it advisable to require that the performance of the variational-difference method is such as to retain the basic advantages of the usual finite-difference methods. As far as we can judge, there are three such advantages (at lease for the finite-difference schemes for elliptic equations):

1. The unknown values of the finite-difference system can be easily related to approximate values of the unknown function of the problem (and, possibly, to some of its derivatives) at grid points.

2. The finite-difference matrix is sparse, that is it contains comparatively few elements which are non-zero. Namely, if N is the order of the finite-difference matrix, the number of its non-zero elements is of order $O(N)$, whereas the total number of matrix elements is N^2.

3. The elements of the finite-difference matrix, that is the coefficients of the finite-difference system of equations as well as the right hand sides of this system, can be calculated comparatively easily.

As an illustration, we consider examples of some simple finite-difference schemes. These schemes were studied in detail in the monographs of Babuška, Vitásek and Práger [1], Forsythe and Wasow [1], and Gavurin [1]. Below, we denote by h the grid size; by j the grid point with abscissa jh (the one-dimensional case); by (j, k) the grid-point with coordinates (jh, kh) (the two-dimensional case). We set $h = 1/n$, where n is an integer. We construct finite-difference schemes for grid-points away from the boundary (finite-difference schemes constructed for grid points in the "boundary layers" usually involve less term).

1. For the linear second order ordinary differential equation

$$p(x)\frac{d^2u}{dx^2} + q(x)\frac{du}{dx} + r(x)u = f(x) \ , \quad 0 < x < 1 \ , \tag{1}$$

the simplest finite-difference scheme is (we only construct the scheme for internal grid-points)

$$p(x_j)|u_h(x_{j+1})-2u_h(x_j)+u_h(x_{j-1})| + hq(x_j)|u_h(x_{j+1})-u_h(x_j)| + \\ + h^2r(x_j)u_h(x_j) = h^2f(x_j) \ . \tag{2}$$

The unknowns of (2) correspond to approximations of the values of the unknown function $u(x)$ at the grid-points (property 1). Each equation contains three unknowns; with $N = n + 1$, the number of non-zero elements of the matrix is equal to $3N + o(N)$ (property 2). Finally, in order to construct (2) one has only to calculate the values of the given functions at grid-points and then do a few simple arithmetic operations (property 3).

2. One of the simplest finite-difference systems for the fourth order ordinary differential equation

$$\frac{d^2}{dx^2}\left(p(x)\,\frac{d^2u}{dx^2}\right) + r(x)u = f(x)\ ,\quad 0 < x < 1\ , \tag{3}$$

is given by (we only construct the scheme for internal grid points)

$$p(x_{j-1})u_h(x_{j-2}) - 2[p(x_{j-1})+p(x_j)]u_h(x_{j-1}) + [p(x_{j-1})+4p(x_j)+p(x_{j+1})]u_h(x_j) - $$
$$- 2[p(x_j)+p(x_{j+1})]u_h(x_{j+1}) + h^4r(x_j)u_h(x_j) = h^4f(x_j)\ . \tag{4}$$

Clearly, the properties 1-3 hold for this scheme. In particular, the number of non-zero matrix elements is $5N + o(N)$.

3. The standard five-point scheme for the Laplace operator of (2.1) yields the following scheme defined on internal grid points

$$4u_{jk} - u_{j-1,k} - u_{j+1,k} - u_{j,k-1} - u_{j,k+1} = h^2f_{jk}\ , \tag{5}$$

where $f_{jk} = f(jh,\ kh)$ and $u_{jk} = u_h(jh,\ kh)$, with $u(jh,\ kh)$ denoting the approximations to the values of $u(jh,\ kh)$. Clearly the number of non-zero matrix elements is $5N + o(N)$.

4. A finite-difference scheme for the two-dimensional biharmonic equation $\Delta^2u = f(x,\ y)$ can be constructed so that the difference equation corresponding to the internal grid point with label $(j,\ k)$ contains 13 unknowns; that is, the approximate values of the solution at the grid points indicated in Figure 9.

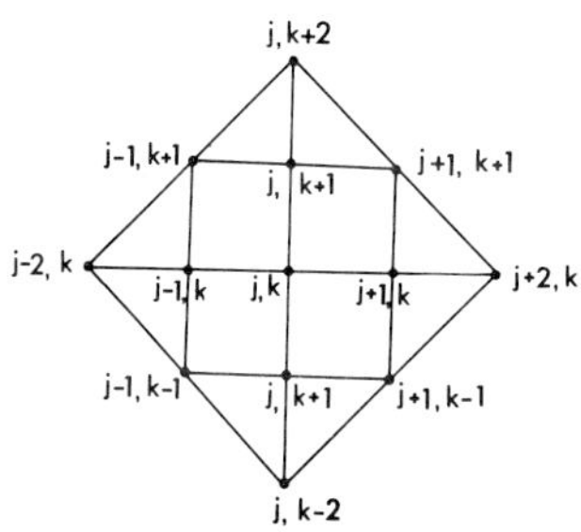

Fig. 9

In fact, the difference equation takes the form

$$20u_{jk} - 8\left(u_{j+1,k}+u_{j-1,k}+u_{j,k-1}+u_{j,k+1}\right) + 2\left(u_{j+1,k+1}+u_{j-1,k+1}+u_{j+1,k-1}+u_{j-1,k-1}\right) +$$
$$+ \left(u_{j+2,k}+u_{j-2,k}+u_{j,k+2}+u_{j,k-2}\right) = h^4f_{jk} \tag{6}$$

where f_{jk} and u_{jk} are defined above. Obviously, the properties 1-3 continue to hold. In particular, the number of non-zero matrix elements equals $13N + o(N)$, where N is the number of grid points.

§4. Primitive Functions and Coordinate Functions

In the present section, we describe methods for constructing coordinate functions for the variational-difference method, for which the advantages of the usual finite-difference method, mentioned in §3, are retained (at least in the author's opinion). It appears that the first such method was proposed by Goël [1] in 1968 for one and two-dimensional problems. This method, but for substantially more general problems, was examined in a number of papers by Strang and Fix on the one hand, and by Mikhlin on the other. In their paper of 1969, Strang and Fix developed this method for second order differential equations with two independent variables. In Mikhlin's paper [7], written in 1970 and published in 1971, this method is developed for arbitrary order differential equations with one variable. These results are presented in detail in Mikhlin [10]. In Strang and Fix's preprint of [2], published at the end of 1970 or at the beginning of 1971, and Mikhlin [8, 9], published in 1971, the same method is developed for equations of arbitrary order and with arbitrarily many variables. In Strang and Fix [2], the method is examined in the Sobolev space W_2^s; while in Mikhlin [8, 9] it is examined in the space W_p^s for arbitrary p, $1 \le p \le \infty$. Additional results about these methods were obtained by Strang [1] and by Demjanovic and Mikhlin [1].

We now introduce some notation and definitions. We denote by $\mathbb{R}_m$ the Euclidian m-dimensional space. We do not distinguish between a point in $\mathbb{R}_m$ and its corresponding vector with center at the origin. In $\mathbb{R}_m$, we work with a fixed cartesian coordinate system. If x is a point in $\mathbb{R}_m$, then its cartesian components will be denoted by $x_1, x_2, \ldots, x_m$ and we shall write $x = (x_1, \ldots, x_m)$. The vector with non-negative integer components in $\mathbb{R}_m$ is called an m-tuple *multi-index*. Below we omit the words "m-tupled". The number $|\alpha| = \alpha_1 + \alpha_2 + \ldots + \alpha_m$ is called the *length* of the multi-index $\alpha = (\alpha_1, \alpha_2, \ldots, \alpha_m)$.

For x a point in $\mathbb{R}_m$, a a constant, α a multi-index and $u(x)$ a function, we use the notation

$$u^{(\alpha)}(x) = D^{\alpha}u(x) = \frac{\partial^{|\alpha|}u(x)}{\partial x_1^{\alpha_1}\partial x_2^{\alpha_2}\ldots\partial x_m^{\alpha_m}}. \tag{1}$$

In addition, $x^{\alpha} = \prod_{i=1}^{m} x_i^{\alpha_i}$.

Let ξ and η be arbitrary m-tuple vectors or multi-indices. We write $\xi \le \eta$, or $\eta \ge \xi$, if $\xi_k \le \eta_k$, $1 \le k \le m$. We also write $\xi < \eta$, or $\eta > \xi$, if $\xi \le \eta$ and, for al least one k , $1 \le k \le m$, $\xi_k < \eta_k$.

The vector with all its components equal to the same number a will be denoted by $\underline{a}$. In the multi-index notation, we then have $\underline{a}^{\alpha} = a^{|\alpha|}$. We shall denote by i a vector with integer components such that $\underline{0} \le i \le \underline{1}$. Each of its components is either zero, or unity. The set I of vectors i $(2^m$ in number$)$ can be identified with the set of vertices $\underline{0} \le t \le \underline{1}$ of the unit cube in $\mathbb{R}_m$.

For a given positive $h \in \mathbb{R}$, we construct a rectangular grid with grid (mesh) size $2h$ in the space R_m such that its sides are parallel to the coordinate axes. We subdivide each cube of the grid (the "*larger*" cubes) to obtain 2^m "*small*" cubes with sides of length h by passing planes parallel to the coordinate planes through the center of the larger cubes. Let the origin coincide with one of the mesh points, then the vectors defined by the grid points have the form jh , where $j = (j_1, \ldots, j_m)$. The symbol jh will be referred to as the *label* of the grid points.

Among all the vertices of a given cube (small or larger), there is one with the smallest magnitude label j . It will be referred to as the *lower corner point* of this cube.

We consider a bounded domain Ω in $\mathbb{R}_m$, and denote by $\hat{\Omega}^h$ the union of all the larger cubes of the rectangular grid which lie inside Ω as shown in Figure 1, and by Ω^h the union of all the larger cubes of the grid which intersect Ω as shown in Figure 10a. Then, we denote by $\hat{\mathcal{J}}^h$ and $\mathcal{J}^h$ the set of labels of the lower corner points of the larger cubes which lie (with overlapping) in $\hat{\Omega}^h$ and Ω^h , respectively; and by $\hat{\mathcal{J}}^h_0$, and $\mathcal{J}^h_0$, the set of labels of the lower corner points of the small cubes which lie in $\hat{\Omega}^h$, and which intersect Ω , respectively. Clearly, $\mathcal{J}^h_0 \subset \mathcal{J}^h$ and $\hat{\mathcal{J}}^h_0 \supset \hat{\mathcal{J}}^h$ as can be seen from Figures 10a and b.

We now define the primitive functions. We limit attention to the simplest class of such functions. More complicated classes will be considered in other chapters, but the complication will be mostly in their (unambiguous) description and not necessary in their construction.

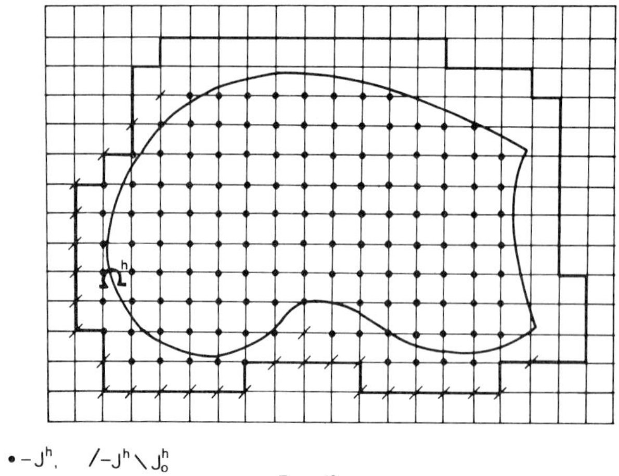

Fig. 10 a

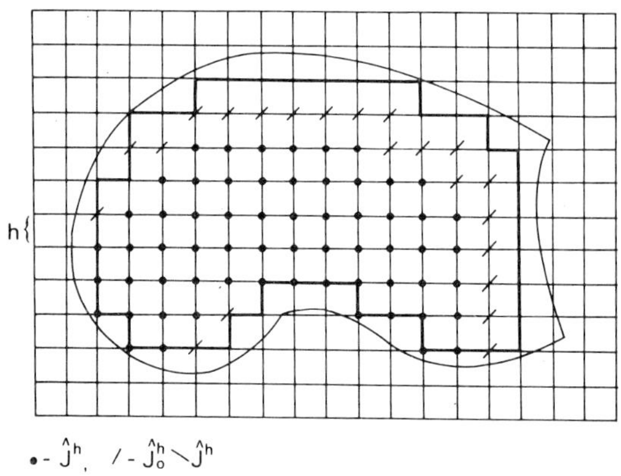

Fig. 10 b

We consider functions $\omega_q(t)$, where t is a point in $\mathbb{R}_m$ and q is a multi-index satisfying the inequality $|q| \le s-1$, with s an integer. We require that $\omega_q(t)$ satisfy the following conditions:

1). $\omega_q \in C^{(s-1)}(\mathbb{R}_m) \cap W_p^{(s)}(\mathbb{R}_m)$, where p is a fixed number, $1 \le p \le \infty$;

2). $\operatorname{supp} \omega_q \subset \tilde{Q}$, $\tilde{Q} = \{t : \underline{0} \le t \le \underline{2}\}$;

3). $\omega_q^{(\alpha)}(\underline{1}) = \delta_{\alpha q}$, $|\alpha|, |q| \le s-1$.

Functions $\omega_q(t)$ which satisfy these three conditions are called *primitive functions*. The set of primitive functions is called *the primitive system of functions*. The numbers m and $s-1$ are called *the dimension* and the *degree* of the primitive system.

Conditions 1). and 2). imply that, on the boundary of the cube $\tilde{Q}$, the boundary conditions are

$$\omega_q^{(\gamma)}(t) = 0 \ , \quad \forall t \in \partial\tilde{Q} \ , \quad |\gamma| \le s-1 \ . \tag{2}$$

We now indicate how the primitive functions can be used to construct variational-difference methods. For example, consider the construction of an approximation to the solution of the first boundary value problem for the following partial differential equation of order $2s$

$$\sum_{|\alpha|=|\beta|\le s} (-1)^{\alpha} D^{\alpha}\left(A_{\alpha\beta}(x) D^{\beta} u\right) = f(x) \ , \quad x \in \Omega \ ,$$

$$u^{(\gamma)}(x) = 0 \ , \quad \forall x \in \partial\Omega \ , \quad |\gamma| \le s-1 \ . \tag{3}$$

We assume that the operator defined by (3) is positive definite in $L_2(\Omega)$, and introduce the following coordinate functions for the application of the variational-difference method to (3):

$$\varphi_{qj}^{(h)}(x) = \omega_q\left(\frac{x}{h} - j\right) \ , \tag{4}$$

where j denotes an arbitrary integer (indicator) vector,and the primitive functions $\omega_q(t)$ are assumed to satisfy condition 1). with $p = 2$. As the required approximation to the solution of (3), we take the following linear combination of coordinate functions

$$u_h(x) = \sum_{|q|=0}^{s-1} \sum_{j\in\hat{J}^h} a_{qj}\varphi_{qj}^{(h)}(x) = \sum_{|q|=0}^{s-1} \sum_{j\in\hat{J}^h} a_{qj}\omega_q\left(\frac{x}{h} - j\right) \ . \tag{5}$$

The corresponding variational-difference system of equations becomes

$$\sum_{|q'|=0}^{s-1} \sum_{j'\in\hat{J}^h} \left[\varphi_{q'j'}^{(h')}, \varphi_{qj}^{(h)}\right] a_{q'j'} = \left(f, \varphi_{qj}^{(h)}\right) \ , \quad \forall q \ , \quad |q| \le s-1 \ , \quad j \in \hat{J}^h \ , \tag{6}$$

where the round brackets denote the scalar product in $L_2(\Omega)$, and the square brackets the scalar product in the energy space (see Mikhlin [3]) of the operator defined by (3).

We now examine the relationships (4)-(6) in detail. Initially, we note that with $j \in \hat{J}^h$, the coordinate functions (4) belong to the energy space of the operator defined by (3). Indeed, for rather general domains Ω, functions from the mentioned energy space must satisfy the following two conditions:

(a) they belong to the space $W_2^{(s)}(\Omega)$;

(b) they satisfy the boundary conditions given in (3).

Condition (a) holds for the functions (4) since the primitive function satisfies condition 1). with $p = 2$. In addition, we have $\hat{\Omega}^h \subset \Omega$, and $j \in \hat{J}^h$ as the label of the lower corner point of one of the larger cubes, which lies in $\hat{\Omega}^h$, and this larger cube is the support of the function $\varphi_{qj}^{(h)}(x)$. Hence, it follows from (2) that condition (b) also holds.

We now choose a finite set $\bar{J}$ of (integer) labels and examine the function

$$v(x) = \sum_{|q|=0}^{s-1} \sum_{j\in\bar{J}} b_{qj}\varphi_{qj}^{(h)}(x) = \sum_{|q|=0}^{s-1} \sum_{j\in J} b_{qj}\omega_q\left(\frac{x}{h} - j\right) . \tag{7}$$

Let α be a multi-index such that $|\alpha| \leq s-1$. We now apply the differentiation operator D^α to both sides of (7) and set $x = \left(j_0+\underline{1}\right)h$, where j_0 is an arbitrary label from the set $\bar{J}$. Using conditions 1).-3). and the identity (2), we obtain

$$b_{\alpha j_0} = h^\alpha v^{(\alpha)}\left(\left(j_0+\underline{1}\right)h\right) \tag{8}$$

and hence,

$$v(x) = \sum_{|q|=0}^{s-1} \sum_{j\in\bar{J}} h^q v^{(q)}\left((j+\underline{1})h\right)\varphi_{qj}^{(h)}(x) = \sum_{|q|=0}^{s-1} \sum_{j\in\bar{J}} h^q v^{(q)}\left((j+\underline{1})h\right)\omega_q\left(\frac{x}{h} - j\right) . \tag{9}$$

If $v(x) \equiv 0$, then we obtain from (8) that all the coefficients $b_{qj} \equiv 0$. Hence, it follows that, for fixed h , the coordinate functions (4) are linearly independent.

The completeness of the system (4) will be examined in Chapter II.

We now demonstrate that, for these chosen coordinate functions and the corresponding form of the approximation (equation (5)), the advantages of the standard finite-difference methods, as mentioned in §3, are usually retained.

Substitution of $u_h(x)$ in (9) yields

$$a_{qj} = h^q u_h^{(q)}\left((j+\underline{1})h\right) . \tag{10}$$

Since $u_h(x)$ is regarded as an approximation to $u(x)$, it follows that

$$a_{qj} \approx h^q u^{(q)}\left((j+\underline{1})h\right) . \tag{11}$$

As yet, we do not examine the question of accuracy. Equation (11) indicates that the unknowns of the variational-difference system (6) have a rather simple representation in terms of the approximate values of the required function and its derivatives at the grid points. Further, the elements of the matrix (6), namely $\left[\varphi^{(n)}_{q'j'}, \varphi^{(n)}_{qj}\right]$, are non-zero only when the larger cubes, with lower corner point labels j' and j, intersect; that is, when $j'_k - j_k = -1, 0, 1$, $1 \leq k \leq m$.

So, for every j, there exist no more than 3^m of the larger cubes which intersect the larger cube with label j. It is not difficult to calculate that the number of multi-indices q, satisfying the inequality $|q| \leq s-1$, equals $M_s = (s+m-1)!/(s-1)!m!$. Consequently the number of non-zero elements in any one of the rows of the matrix (6) does not exceed $3^m M_s$.

If N is the number of grid points in $\hat{\Omega}^h$, then, among the $M_s N^2$ elements of the matrix (6), there are not more than $3^m M_s N$ non-zero ones. Thus, the matrix (6) will be rather sparse once $N >> 3^m$.

The non-zero elements of the variational-difference matrix (6) are more difficult to calculate than for a standard finite-difference matrix. However, it should be noted that, when the coefficients of (3) are constant, the calculation of the matrix (6) reduces to the calculation of a fixed number of integrals of products of the primitive functions with themselves and their derivatives.

In Chapter VIII, we shall present a method which enables the calculation of the matrix (6) to be reduced to the approximate calculation of a fixed number of standard integrals when the coefficients $A_{\alpha\beta}$ are variable, but sufficiently smooth. Thus, we can conclude that property (3) will in general hold for the variational-difference method.

If the differential equation (3) is examined with natural boundary conditions, then the above analysis of (3) will in general remain valid once $\hat{\Omega}^h$ is replaced by Ω^h. However, it should be noted that there is now one difficulty associated with the construction of (6). It is the calculation of integrals along the boundary $\partial\Omega$ when it does not coincide with Ω^h (see Figure 10a, where such cubes are shaded).

We conclude this section with the following observation. Sometimes it turns to be more convenient to change slightly the definition of the primitive function; namely, to replace condition (2) by

$$(2) \quad \operatorname{supp} \omega_q \subset \{t : -1 \leq t \leq 1\} .$$

Some subsequent formulas must also be changed. It is necessary to replace condition (3) with

$$(3) \quad \omega_q^{(\alpha)}(\underline{0}) = \delta_{\alpha q} , \quad |\alpha|, |q| \leq s-1 .$$

We can retain (4)-(6) unchanged, as well as the form of the approximate solution and the variational-difference system, but it is necessary to change the definition of the set $\hat{J}^h$. It is now necessary to take for $\hat{J}^h$ the set of labels of the centers of the larger cubes which lie in $\hat{\Omega}^h$. If $v(x)$ is defined by (7), then (8) and (9) must be replaced by

$$b_{\alpha j_0} = h^\alpha v^{(\alpha)}(j_0 h) \tag{8_1}$$

and

$$v(x) = \sum_{|q|=0}^{s-1} \sum_{j\in\overline{J}} h^q v^{(q)}(jh)\omega_q\left(\frac{x}{h} - j\right) . \tag{9_1}$$

§5. Interpolatory Properties of Primitive Systems of Functions

Let $\{\omega_q(t)\}$ be a primitive system of dimension m and degree $s - 1$. Let $u \in C^{(s-1)}(\overline{\Omega})$. Consider the set of integer-vectors J such that $(j+\underline{1})h \in \overline{\Omega}$, $\forall j \in J$, and construct the function

$$\tilde{u}(x) = \sum_{|q|=0}^{s-1} \sum_{j\in J} h^q u^{(q)}\big((j+\underline{1})h\big)\omega_q\left(\frac{x}{h} - j\right) . \tag{1}$$

From (4.7) and (4.8) it follows that

$$u^{(q)}\big((j+1)h\big) = \tilde{u}^{(q)}\big((j+\underline{1})h\big) \ , \quad \forall q \ , \quad |q| \le s-1 \ , \quad \forall j \in y \ .$$

Consequently, the function $\tilde{u}(x)$ yields the solution of the following problem of Hermite interpolation: construct a function $\tilde{u}(x)$ which along with all its derivatives up to order $s - 1$ coincides with the function $u(x)$ and its derivatives at given points $(j+\underline{1})h$, $j \in J$.

It can happen that, at certain points $(j_0+\underline{1})h$, not all the mentioned derivatives of $u(x)$ are given. For each vector $j \in J$, introduce a set M_j of non-equal multi-indices $q : M_j = \left(q^{(j_1)}, q^{(j_2)} \ldots\right)$, where $q^{(j_k)} \ne q^{(j_l)}$, $k \ne l$, and $|q^{(j_k)}| \le s-1$. We require the function $\tilde{u}(x)$ defined by the conditions

$$\tilde{u}^{(q)}\big((j+\underline{1})h\big) = u^{(q)}\big((j+\underline{1})h\big) \ ; \quad \forall j \in J \ , \quad \forall q \in M_j \ . \tag{3}$$

Certainly, the function (1) is a solution of this problem, if we construct a solution which satisfies only a subset of the conditions:

$$\tilde{u}(x) = \sum_{j\in J} \sum_{q\in M_j} h^q u^{(q)}\big((j+1)h\big)\omega_q\left(\frac{x}{h} - j\right) . \tag{4}$$

The question of the accuracy of such an interpolation formula will be examined in Chapters II and III, and for more special circumstances in Chapters V and VI.

CHAPTER II

COMPLETENESS AND FUNDAMENTAL COMPLETENESS CONDITIONS

The coordinate functions, constructed in §4 of the previous chapter, do not in general form a complete system. The aim of the present chapter is to formulate additional conditions which guarantee that a given coordinate system is complete with respect to some chosen topology. We refer to such conditions as *fundamental completeness conditions.*

§1. Approximation of Smooth Functions

Let $u \in \overset{\circ}{W}{}_p^{(s)}(\Omega)$, where $\overset{\circ}{W}{}_p^{(s)}(\Omega)$ denotes the set of functions of $W_p^{(s)}(\Omega)$ which satisfy the conditions

$$u^{(\gamma)}(x)|_{\partial\Omega} = 0 \ , \quad |\gamma| \le s-1 \ , \tag{1}$$

on the boundary $\partial\Omega$ of the region Ω . Let $1 \le p \le \infty$ and let Ω be a *region of Sobolev type*: the union of a finite number of regions, each of which is starshaped with respect to every point of some ball the choice of which depends on each region. Consider the primitive system $\{\omega_q(t)\}$, satisfying the conditions 1).-3). of §4 in Chapter I, and find conditions under which the coordinate system (4.4) of Chapter I is complete in $\overset{\circ}{W}{}_p^{(s)}(\Omega)$.

We denote by $C_0^{(s)}(\overline{\Omega})$ the set of functions which are continuous, along with their derivatives up to order s , in the closed region $\overline{\Omega} = \Omega \cup \partial\Omega$ and satisfy the boundary conditions (1). In this section, we limit attention to the situation when the required solution $u \in C_0^{(s)}(\overline{\Omega})$. As a first step, it is useful to obtain fundamental completeness conditions for this situation. We consider the region $\hat{\Omega}^h$ and construct the approximation $u^h(x)$ using (5.1) of Chapter I with $J = \hat{J}^h$,

$$u^h(x) = \sum_{|q|=0}^{s-1} \sum_{j\in\hat{J}^h} h^q u^{(q)}\big((j+1)h\big)\omega_q\left(\frac{x}{h} - j\right) . \tag{2}$$

Since $u^h(x)$ can be considered as an approximation to $u(x)$, we estimate the norm of the difference $u - u^h$ in $\overset{\circ}{W}{}^{(s)}_p(\Omega)$,

$$\|u-u^h\|_{\overset{\circ}{W}{}^{(s)}_p(\Omega)} = \left\{\int_\Omega \sum_{|\alpha|=s} |D^\alpha u(x) - D^\alpha u^h(x)|^p dx\right\}^{1/p} . \tag{3}$$

We are interested in conditions under which (3) tends to zero as h tends to zero. Since $u^h(x) \equiv 0$ in the region $\Omega\backslash\hat{\Omega}^h$, the corresponding integrals in (3) have the form

$$\int_{\Omega\backslash\hat{\Omega}^h} |u^{(\alpha)}(x)|^p dx , \quad |\alpha| = s . \tag{4}$$

We now assume (and this assumption is retained below) that the measure of the region $\Omega\backslash\hat{\Omega}^h$ tends to zero as $h \to 0$. As a direct consequence, (4) tends to zero as h tends to zero, so it is sufficient to examine

$$\lim_{h\to 0} \int_{\hat{\Omega}^h} \sum_{|\alpha|=s} \left|D^\alpha\left[u(x) - \sum_{|q|=0}^{s-1} \sum_{j\in\hat{J}^h} h^q u^{(q)}\big((j+\underline{1})h\big)\omega_q\left(\frac{x}{h} - j\right)\right]\right|^p dx = 0 . \tag{5}$$

We denote by Q_j that small cube of the grid which has lower corner vertex jh . In this cube, $jh \le x \le (j+\underline{1})h$. Thus, (5) holds, if

$$\int_{Q_{j_0}} \left|D^\alpha\left[u(x) - \sum_{|q|=0}^{s-1} \sum_{j\in\hat{J}^h} h^q u^{(q)}\big((j+\underline{1})h\big)\omega_q\left(\frac{x}{h} - j\right)\right]\right|^p dx = o\big(h^m\big) ;$$

$$\forall j_0 \in \hat{J}^h , \quad \forall\alpha \text{ s.t. } |\alpha| = s . \tag{6}$$

We identify the primitive functions $\omega_q\left(\frac{x}{h} - j\right)$ which are not identically equal to zero on the cube Q_{j_0} . This will occur if and only if $\underline{0} \le \frac{x}{h} - j \le \underline{2}$. But x is in Q_{j_0} , so $x = \big(j_0+t\big)h$, where $\underline{0} \le t \le \underline{1}$; hence, we have $\underline{0} \le j_0-j+t \le \underline{2}$ for any t , $\underline{0} \le t \le \underline{1}$. This will hold if and only if $\underline{0} \le j_0-j \le \underline{1}$. The last inequality implies that $j_0 - j = i$, $i \in I$ (see §4 of Chapter I). Substitution of $x = x_0 + th$ and $x_0 = j_0h$ into (6) yields

$$\int_Q \left| u^{(\alpha)}(x_0+th) - \sum_{|q|=0}^{s-1} \sum_{i\in I} h^{|q|-|\alpha|} u^{(q)}\big(x_0+(\underline{1}-i)h\big)\omega_q^{(\alpha)}(t+i) \right|^p dt = o(1) ,$$

$$\forall\alpha \text{ s.t. } |\alpha| = s , \quad (7)$$

where Q is the cube $\underline{0} \le t \le \underline{1}$.

The equality (7) remains valid, if we remove any term from the integrand which tends uniformly to zero as h tends to zero. Thus, we can replace $u^{(\alpha)}(x_0+th)$ by $u^{(\alpha)}(x_0)$. We therefore consider the Taylor series expansion for $u^{(q)}\big(x_0+(\underline{1}-i)h\big)$ up to terms of order $h^{s-|q|}$,

$$u^{(q)}\big(x_0+(\underline{1}-i)h\big) = \sum_{|\beta|=0}^{s-|q|} \frac{h^\beta}{\beta!} (\underline{1}-i)^\beta u^{(q+\beta)}(x_0) + \rho_{s-|q|}$$

and discard the higher order terms. Instead of (7), we obtain

$$\int_Q \left| u^{(\alpha)}(x_0) - \sum_{|q|=0}^{s-1} \sum_{|\beta|=0}^{s-|q|} \sum_{i\in I} \frac{h^{|\beta+q|-s}}{\beta!} (\underline{1}-i)^\beta u^{(\beta+q)}(x_0)\omega_q^{(\alpha)}(t+i) \right|^p dt = o(1) ,$$

$$\forall\alpha \text{ s.t. } |\alpha| = s . \quad (8)$$

We now transform the inner sum. Initially, we pick out the terms which do not contain h . They correspond to $|\beta| = s - |q|$ and form the following sum ($\gamma = \beta + q$)

$$\sum_{|\gamma|=s} u^{(\gamma)}(x_0) \sum_{\substack{|q|=0\\ q\le\gamma}}^{s-1} \sum_{i\in I} \frac{(\underline{1}-i)^{\gamma-q}}{(\gamma-q)!} \omega_q^{(\alpha)}(t+i) .$$

The remaining terms have the form

$$\sum_{|q|=0}^{s-1} \sum_{|\beta|=0}^{s-|q|-1} \sum_{i\in I} \frac{h^{|\beta+q|-s}}{\beta!} (\underline{1}-i)^\beta u^{(\beta+q)}(x_0)\omega_q^{(\alpha)}(t+i) .$$

Setting $\beta + q = \gamma$ in order to eliminate the summation with respect to β and changing the order of the summation, we obtain

$$\sum_{|\gamma|=0}^{s-1} h^{|\gamma|-s} u^{(\gamma)}(x_0) \sum_{q\le\gamma} \sum_{i\in I} \frac{(\underline{1}-i)^{\gamma-q}}{(\gamma-q)!} \omega_q^{(\alpha)}(t+i) .$$

Using this in (8) and taking account of the fact that

$$u^{(\alpha)}(x_0) = \sum_{|\gamma|=s} \delta_{\alpha\gamma} u^{(\gamma)}(x_0) , \quad |\alpha| = s ,$$

we then obtain

$$\int_Q \Big| \sum_{|\gamma|=s} \Big[\delta_{\alpha\gamma} - \sum_{|q|=0}^{s-1} \sum_{i\in I} \frac{(\underline{1}-i)^{\gamma-q}}{(\gamma-q)!}\, \omega_q^{(\alpha)}(t+i)\Big] u^{(\gamma)}(x_0) - $$

$$- \sum_{|\gamma|=0}^{s-1} h^{|\gamma|-s} u^{(\gamma)}(x_0) \sum_{q\le\gamma} \sum_{i\in I} \frac{(\underline{1}-i)^{\gamma-q}}{(\gamma-q)!}\, \omega_q^{(\alpha)}(t+i)\Big|^p dt = o(1) \;, \quad \forall\alpha \text{ s.t. } |\alpha| = s \;. \tag{9}$$

The equality (9) is valid if and only if the following conditions hold:

$$\sum_{\substack{|q|=0\\ q\le\gamma}}^{s-1} \sum_{i\in I} \frac{(\underline{1}-i)^{\gamma-q}}{(\gamma-q)!}\, \omega_q^{(\alpha)}(t+i) = \delta_{\alpha\gamma}, \quad |\alpha| = |\gamma| = s \;,$$

$$\sum_{q\le\gamma} \sum_{i\in I} \frac{(\underline{1}-i)^{\gamma-q}}{(\gamma-q)!}\, \omega_q^{(\alpha)}(t+i) = 0 \quad, \quad |\alpha| = s \;, \quad |\gamma| < s \;, \quad \underline{0} \le t \le \underline{1} \;.$$

After integration, we obtain

$$\sum_{\substack{|q|=0\\ q\le\gamma}}^{s-1} \sum_{i\in I} \frac{(\underline{1}-i)^{\gamma-q}}{(\gamma-q)!}\, \omega_q(t+i) = \frac{t^\gamma}{\gamma!} + \varphi_\gamma(t) \;, \quad |\gamma| = s \;, \tag{10a}$$

$$\sum_{q\le\gamma} \sum_{i\in I} \frac{(\underline{1}-i)^{\gamma-q}}{(\gamma-q)!}\, \omega_q(t+i) = \psi_\gamma(t) \;, \quad |\gamma| < s \;, \tag{10b}$$

where $\underline{0} \le t \le \underline{1}$, and $\varphi_\gamma(t)$ and $\psi_\gamma(t)$ are polynomials of degree less than or equal to $s - 1$.

We now prove that $\varphi_\gamma(t) \equiv 0$. We differentiate the identity (10a) using the operator D^β , $|\beta| \le s-1$ and then set $t = \underline{0}$. If $i \ne \underline{1}$, then at least one of the components of the vector i is zero and because of the boundary conditions (4.2) of Chapter I it follows that $\omega_q^{(\beta)}(i) = 0$. If $i = \underline{1}$ then, since $|\gamma| = 1$ and $|q| \le s-1$, $\gamma-q > 0$ and $(\underline{1}-i)^{\gamma-q} = 0$. Thus, we have zero on the left hand side and $|D^\beta t^\gamma/\gamma!|_{t=0} = 0$ on the right hand side. Finally, since $\varphi_\gamma^{(\beta)}(\underline{0}) = 0$, for all $|\beta|$, $|\beta| \le s-1$, it follows that $\varphi_\gamma(t) \equiv 0$.

We now prove that $\psi_\gamma(t) = t^\gamma/\gamma!$ for all γ , $|\gamma| < s$. We apply the operator D^β , $|\beta| < s$, to (10b) and set $t = \underline{0}$. By condition 3). of §4 of Chapter I, we obtain on the left hand side the value $\delta_{\beta\gamma}$. Thus $\psi_\gamma^\beta(\underline{0}) = \delta_{\beta\gamma}$ and, consequently, $\psi_\gamma(t) = t^\gamma/\gamma!$.

We can combine (10a) and (10b) by writing them as

$$\sum_{\substack{|q|=0 \\ q\leq\gamma}}^{s-1} \sum_{i\in I} \frac{(\underline{1}-i)^{\gamma-q}}{(\gamma-q)!} \omega_q(t+i) = \frac{t^\gamma}{\gamma!}, \quad \underline{0} \leq t \leq \underline{1}, \quad |\gamma| \leq s . \tag{11}$$

These are the required *fundamental completeness conditions.* They yield necessary and sufficient conditions for a function $u \in C_0^{(s)}(\bar{\Omega})$ to be approximated by $u^h(x)$ in $\overset{\circ}{W}{}^s_p(\Omega)$, when $u^h(x)$ is defined by (2). It follows naturally that $u^h(x)$ approximates $u(x)$ in $C^{(s-1)}(K)$, where K denotes any compact subregion of Ω.

§2. Extensions of Functions

Let G and G_1 be bounded or unbounded regions in $\mathbb{R}_m$ with $G \subset G_1$. We say that $W_p^{(s)}(G)$ has been *extended to* G_1, if for each function $u \in W_p^{(s)}(G)$, a function $u^*(x)$ with the following properties can be identified with it:

(a) $u^* \in W_p^{(s)}(G_1)$;

(b) $\forall x \in G$, $u^*(x) = u(x)$;

(c) there exists a constant C such that

$$\forall u \in W_p^{(s)}(G), \quad \|u^*\|_{W_p^{(s)}(G_1)} \leq C\|u\|_{W_p^{(s)}(G)} . \tag{1}$$

We say that $u^*(x)$ is a *structure-preserving extension of* $u(x)$ to the region G_1.

The notion of an extension, which preserves structure, can obviously be applied to other spaces of functions than $W_p^{(s)}$. The initial work on the extension of functions was applied to the space $C^{(s)}$. This was the classical work of Whitney and Hestenes.

If the region G_1 is unbounded, then we can assume that the function being extended equals zero outside some ball. To achieve this, it suffices to multiply by any function from $C^\infty(\mathbb{R}_m)$ which equals unity in G and vanishes outside the mentioned ball.

We again consider the space $W_p^{(s)}$. We are now interested in the situation where $G = \Omega$ is a bounded region in $\mathbb{R}_m$ and $G_1 = \mathbb{R}_m$. It is necessary to add one more condition to (a)-(c):

(d) the support of $u^*(x)$ is contained in some ball, independent of u.

The most general theorem on the extension of $W_p^{(s)}(\Omega)$ to $\mathbb{R}_m$ is due to Calderon [1]. It will be stated and proved below.

We say that a surface $\Gamma \subset \mathbb{R}_m$ is Lipschitz and write $\Gamma \in C^{(0,1)}$, if it satisfies the following properties: for any $x \in \Gamma$ a coordinate system can be constructed with origin at x such that that part of the surface Γ, which lies in some neighbourhood of x, has an equation of the form

$$\xi_m = f(\overline{\xi}) \ , \quad \overline{\xi} = (\xi_1, \xi_2, \ldots, \xi_{m-1}) \ , \tag{2}$$

where $f(\overline{\xi})$ satisfies the Lipschitz condition

$$|f(\overline{\xi}') - f(\overline{\xi}'')| \leq A|\overline{\xi}' - \overline{\xi}''| \ , \tag{3}$$

with A a constant independent of x.

THEOREM (Calderon [1]). *Let Ω be a bounded region in $\mathbb{R}_m$ with $\partial\Omega \in C^{(0,1)}$. Let s denote a positive integer, and $p \in \mathbb{R}$ satisfy $1 < p < \infty$. Then there exists an extension of $W_p^{(s)}(\Omega)$ to $\mathbb{R}_m$ which satisfies conditions* (a)-(d).

We divide the proof into three parts:

1°. Consider initially the case when Ω is starshaped with respect to some given ball. Let $u \in W_p^{(s)}(\Omega)$. The Sobolev integral representation for u (see Sobolev [1] and Mikhlin [1]) becomes

$$u(x) = \sum_{|\alpha|=0}^{s-1} x^\alpha \int_K b_\alpha(y)u(y)dy + \sum_{|\alpha|=s} \int_\Omega \frac{A_\alpha(x,y)}{r^{m-s}} u^{(\alpha)}(y)dy \ ,$$
$$x \in \Omega \ , \quad r = |x-y| \ . \tag{4}$$

Here K is the ball with respect to which Ω is starshaped, and the $b_\alpha(y)$ are infinitely differentiable functions. Then the functions $A_\alpha(x, y)$ are infinitely differentiable with respect to x and y if $x \neq y$. They are bounded and vanish, if y lies outside the region G_x, which is the region bounded by ∂K and the cone with vertex x which is tangential to this sphere (see Figure 11). It can be readily proved that

$$D_x^\beta A_\alpha(x, y) = r^{-|\beta|} A_{\alpha\beta}(x, y) \ , \tag{5}$$

where the functions $A_{\alpha\beta}$ are bounded, and infinitely often differentiable, if $x \neq y$.

Now, for any point $x \in \mathbb{R}_m$, we put

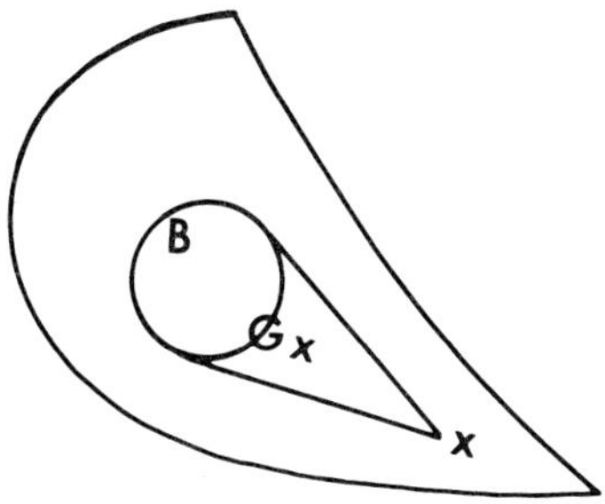

Fig. 11

$$\tilde{u}(x) = \sum_{|\alpha|=0}^{s-1} x^{\alpha} \int_K b_{\alpha}(y)u(y)dy + \sum_{|\alpha|=s} \int_{\Omega} \frac{A_{\alpha}(x,y)}{r^{m-s}} u^{(\alpha)}(y)dy \ . \tag{6}$$

Obviously,

$$\tilde{u}(x) = u(x) \ , \quad \forall x \in \Omega \ , \tag{7}$$

which ensures that condition (b) is satisfied. We now prove that $\tilde{u} \in W_p^{(s)}(\Omega_1)$ for any bounded region $\Omega_1 \subset \mathbb{R}_m$ and that

$$\|\tilde{u}\|_{W_p^{(s)}(\Omega_1)} \le C(\Omega, \Omega_1)\|u\|_{W_p^{(s)}(\Omega)} \ , \tag{8}$$

where C does not depend on u .

Using theory about the differentiation of integrals with a weak singularity (see Mikhlin [1], [4]), the derivatives $\tilde{u}^{(\gamma)}(x)$, $|\gamma| \le s-1$, can be obtained by differentiating (6) under the integral sign. From (5) it follows that, as a result, we obtain integrals with a weak singularity. We can interpret them as integrals on Ω_1 , by putting $u(x) = 0$ when $x \in \Omega_1 \backslash \Omega$. The derivatives of order s of $\tilde{u}(x)$ can be represented by certain singular integrals which can be defined on $\mathbb{R}_m$ by putting $u(x) = 0$ outside Ω . Using Calderon and Zigmund's [1] theorem on the boundedness of singular integral operator in $L_p(\mathbb{R}_m)$, we obtain

$$\sum_{|\alpha|=s} \int_{\mathbb{R}_m} |\tilde{u}^{(\alpha)}(x)|^p dx \le C^p \sum_{|\alpha|=s} \int_{\mathbb{R}_m} |u^{(\alpha)}(x)|^p dx =$$

$$= C^p \sum_{|\alpha|=s} \int_{\Omega} |u^{(\alpha)}(x)|^p dx \le C^p \|u\|^p_{W_p^{(s)}(\Omega)} \ , \quad C = \text{const.} \tag{9}$$

From (6) it also follows that, for any region $\Omega_1 \subset \mathbb{R}_m$,

$$\left\{\int_{\Omega_1} |\tilde{u}(x)|^p dx\right\}^{1/p} \le C(\Omega, \Omega_1)\|u\|_{W_p^{(s)}(\Omega)} . \tag{10}$$

Together (9) and (10) yield that $\tilde{u} \in W_p^{(s)}(\Omega_1)$ and

$$\|\tilde{u}\|_{W_p^{(s)}(\Omega_1)} \le C_0(\Omega, \Omega_1)\|u\|_{W_p^{(s)}(\Omega)} . \tag{11}$$

We take the ball $|x| < 2R$ as the region Ω_1, we choose R so that the region Ω lies inside the ball $|x| \le R$. We denote by $\zeta(x)$ a function of $C^\infty(\mathbb{R}_m)$ such that

$$\zeta(x) = \begin{cases} 1, & |x| \le R, \\ & \qquad 0 < \zeta(x) < 1, \quad R < x < 2R ; \\ 0, & |x| \ge 2R, \end{cases}$$

and put

$$u^*(x) = \tilde{u}(x)\zeta(x) . \tag{12}$$

Obviously, the formula (12) gives the required extension of $u(x)$. In fact, $u^* \in W_p^{(s)}(\mathbb{R}_m)$ and $u^*(x) = u(x)$, $x \in \Omega$, so the conditions (a), (b) are satisfied. The condition (d) also holds, since $u^*(x) = 0$ when $|x| \ge 2R$. Finally,

$$\begin{aligned}\|u^*\|^p_{W_p^{(s)}(\mathbb{R}_m)} = \|u^*\|^p_{W_p^{(s)}(\Omega_1)} &= \int_{\Omega_1} |\tilde{u}(x)\zeta(x)|^p dx + \sum_{|\alpha|=s} \int_{\Omega_1} |D^\alpha(\tilde{u}(x)\zeta(x))|^p dx \\ &\le \int_{\Omega_1} |\tilde{u}(x)|^p dx + C_1 \sum_{|\alpha|\le s} \int_{\Omega_1} |u^{(\alpha)}(x)|^p dx \\ &\le C_2 \|\tilde{u}\|^p_{W_p^{(s)}(\Omega_1)} \le C_3\|u\|^p_{W_p^{(s)}(\Omega)} , \quad C_j = \text{const}, \\ &\qquad j = 1, 2, 3 ,\end{aligned}$$

where C_3 depends only on Ω. Since condition (c) is thereby established, the Calderon theorem is proved for regions which are starshaped with respect to some given ball.

2°. We now drop the assumption that Ω is starshaped, and replace it with the assumption that $\partial\Omega \in C^{(0,1)}$. We choose an arbitrary point x_0 on $\partial\Omega$ and take it as the origin of a coordinate system for which (2) defines the above mentioned part of $\partial\Omega$. We construct the spherical cylinder $|\overline{\xi}| = \delta$ with axis ξ_m and a sufficiently small radius δ. Consider the region Z bounded by the surface $\partial\Omega$, the mentioned cylindrical surface and the plane $\xi_m = C_0 = \text{const}$.

We choose the constant C_0 so small that the region Z lies inside Ω

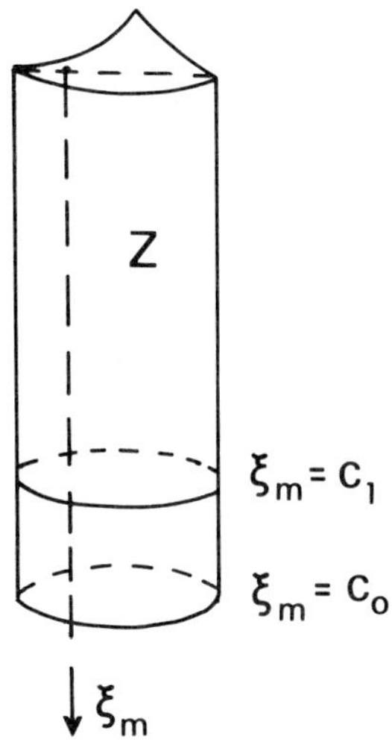

Fig. 12

(Figure 12). We choose the number C_1 such that $0 < C_1 < C_0$ and sufficiently close to C_0 and then draw the plane $\xi_m = C_1$. We prove that the region Z is starshaped with respect to every point, lying between the planes $\xi_m = C_0$ and $\xi_m = C_1$. Let $\xi^{(0)} = \left(\overline{\xi}^{(0)}, \xi_m^{(0)}\right)$ be such a point. We draw a ray through it. If the ray has no common points with $\Gamma' = \partial\Omega \cap \partial Z$, then it is clear that it crosses ∂Z only once. Suppose that the ray crosses Γ' and has two common points with ∂Z. This ray leaves the point $\xi^{(0)}$ such that $\xi_m^{(0)} - f\left(\overline{\xi}^{(0)}\right) > 0$. We prove that, for sufficiently small δ, it crosses Γ' and then appears at that part of $\mathbb{R}_m$ where $\xi_m - f(\overline{\xi}) < 0$. We write the equation of the ray as

$$\frac{\xi_1-\xi_1^{(0)}}{p_2} = \dots = \frac{\xi_{m-1}-\xi_{m-1}^{(0)}}{p_{m-1}} = \frac{\xi_m-\xi_m^{(0)}}{1} = t .$$

If δ is sufficiently small then angular coefficients $p_1, \dots, p_{m-1}$ are arbitrary small. Denoting the $(m-1)$-tupled $\left(p_1, p_2, \dots, p_{m-1}\right)$ by $\overline{p}$, we have

$$\overline{\xi} = \overline{\xi}^{(0)} + t\overline{p} , \quad \xi_m = \xi_m^{(0)} + t . \tag{13}$$

Obviously, since ξ_m decreases, the value t decreases along the ray from $\xi^{(0)}$ to the surface Γ'. Let the ray (13) cross Γ' at the point ξ', corresponding to the value $t = t'$. Then $\xi_m^{(0)} + t' - f(\overline{\xi}^{(0)}+t'\overline{p}) = 0$. We take $t'' < t'$, denote by $\xi'' = (\overline{\xi}'', \xi_m'')$ the corresponding point on the ray (13) and evaluate

$$\xi_m'' - f(\overline{\xi}'') = \xi_m^{(0)} + t'' - f(\overline{\xi}^{(0)}+t''\overline{p}) = \xi_m^{(0)} + t'' - f(\overline{\xi}^{(0)}+t''\overline{p}) - \left(\xi_m^{(0)}+t'-f(\overline{\xi}^{(0)}+t'\overline{p})\right)$$
$$= -\left[t'-t''-f(\overline{\xi}^{(0)}+t'\overline{p})+f(\overline{\xi}^{(0)}+t''\overline{p})\right] .$$

The expression in square brackets is positive for small values of δ. It dominates the value $t' - t'' - A|\overline{p}|(t'-t'')$, where A is the Lipschitz constant of f. The last expression is positive, if $A|\overline{p}| < 1$. Hence it follows that the point ξ'' lies in the region where $\xi_m - f(\overline{\xi}) < 0$ (that is, outside $\overline{Z}$), and the ray (13) cannot cross the boundary of Z again.

From 1°, there exists a structure preserving extension of $u(x)$ outside of Z and, in particular, outside $\Gamma' \in \partial\Omega$. Hence it follows that any point $x \in \partial\Omega$ can be made the center of a ball to which any function $u \in W_p^{(s)}(\Omega)$ can be extended.

3°. On the basis of the Heine-Borel lemma one can select a finite number of balls from the above mentioned set of balls, so that they form a covering of $\partial\Omega$. Let them be denoted by $K_1, K_2, \ldots, K_n$. The union $\bigcup_{j=1}^{n} K_j$ is a neighbourhood of the surface $\partial\Omega$. We construct a region $K_0 \subset \Omega$ such that the regions $K_0, K_1, \ldots, K_n$ form a covering of Ω. We then add the regions $K_{n+1}, K_{n+2}, \ldots$ such that all these regions form a locally finite covering for $\mathbb{R}_m$ (see, for example, Shilov [1]). We select $K_{n+1}, K_{n+2}, \ldots$ so that they do not intersect Ω. Let $\{\varphi_j(x)\}$ be a partition of unity for the covering $\{K_j\}$. Then $\varphi_j(x) = 0$, if $j > n$ and $x \in \Omega$.

We denote by $u_j(x)$ the extension of $u(x)$ over the ball K_j, $1 \le j \le n$, constructed in 2°. We put $u_0(x) = u(x)$ for $x \in K_0$, and let

$$w_j(x) = \begin{cases} u_j(x)\varphi_j(x) , & x \in K_j , \\ 0 , & x \notin K_j , \quad j = 0, 1, \ldots, n . \end{cases}$$

The supports of the functions w_j are compact and $w_j \in W_p^{(s)}(\mathbb{R}_m)$. We now put

$$u^*(x) = \sum_{j=0}^{n} w_j(x) . \tag{14}$$

Obviously, the support of $u^*(x)$ is compact and $u^* \in W_p^{(s)}(\mathbb{R}_m)$. Let $x \in \Omega$, then $u_j(x) = u(x)$ and

$$u^*(x) = u(x) \sum_{j=0}^{n} \varphi_j(x) = u(x) \sum_{j=0}^{\infty} \varphi_j(x) = u(x) .$$

Thus the function $u^*(x)$ gives the required structure-preserving extension of $u(x)$ to the whole space $\mathbb{R}_m$.

The reasoning of the present section is not valid for $p = 1$ or $p = \infty$, since the theorem about the boundedness of the singular integral operator is not valid. But the theorem remains valid for the spaces $W_1^{(s)}(\Omega)$ and $W_\infty^{(s)}(\Omega)$ provided $\partial\Omega \in C^{(s)}$, since the reasoning of Whitney and Hestenes remain valid. For details see Babitch [1], Nikolskiy [1].

§3. Completeness in Sobolev Spaces[1]

Let $u \in W_p^{(s)}(\Omega)$ and Ω be such that the function $u(x)$ can be extended to the whole of $\mathbb{R}_m$ so that the extended function belongs to $W_p^{(s)}(\mathbb{R}_m)$. We denote this extension by $u^*(x)$. The support of $u^*(x)$ is compact, so we construct an open cube Q containing this support. The cube can be chosen so that its sides are parallel to the coordinate axes. Choosing the origin in an appropriate way, Q can be defined by the inequalities $\underline{0} \le x \le \underline{a}$. We average the function $u^*(x)$ taking the averaging radius ρ smaller than the distance between the sets ∂Q and $\operatorname{supp} u^*(x)$. We denote the averaged function by $v_\rho(x)$. It has the following properties (see Sobolev [1])

$$v_\rho \in C^{(\infty)}(\mathbb{R}_m) , \quad \operatorname{supp} v_\rho \subset Q , \quad \|u^*-v_\rho\|_{p,s} \xrightarrow[\rho\to 0]{} 0 , \tag{1}$$

where $\|\cdot\|_{p,s}$ denotes the norm in $W_p^{(s)}(\mathbb{R}_m)$.

On the boundary of Q , $v\ (x)$ satisfies the boundary conditions

$$v_\rho^{(\gamma)}(x)\big|_{\partial Q} = 0 , \quad \forall\gamma .$$

Consequently, this function has the form described in §1 of the present chapter: if the primitive functions satisfy the fundamental completeness conditions, then $v_{\bar\rho}(x)$ can be approximated in $W_p^{(s)}(Q)$ by a linear combinations of the coordinate functions (4.4) of Chapter I. We now take an arbitrary $\varepsilon > 0$, select an averaging radius so that $\|u^*-v_\rho\|_{p,s} < \varepsilon/2$ and fix the value of ρ . According to the results of §1 we

---

[1] See Mikhlin [11].

have, for sufficiently small h , $\left\|v_\rho - v_\rho^{(h)}\right\|_{p,s} < \varepsilon/2$, where $v_\rho^{(h)}$ is constructed from v_ρ using (1.2). From the last two inequalities it follows that

$$\left\|u^* - v_\rho^{(h)}\right\|_{W_p^{(s)}(Q)} < \varepsilon .$$

By discarding the non-negative integral over $Q\backslash\Omega$ and remembering that $u^*(x) = u(x)$, $x \in \Omega$, we obtain

$$\left\|u - v_\rho^{(h)}\right\|_{W_p^{(s)}(\Omega)} < \varepsilon . \tag{2}$$

Thus, any function of $W_p^{(s)}(\Omega)$ can be approximated (with respect to its topology) by a linear combination of the coordinate functions (4.4) of Chapter I, if the primitive functions satisfy the fundamental completeness conditions. Consequently, with respect to such conditions, the system of primitive functions is complete in $W_p^{(s)}(\Omega)$.

We now pose the following question: if we assume that some of the fundamental completeness conditions do not hold, which functions of $W_p^{(s)}(\Omega)$ can be approximated by linear combinations of the corresponding coordinate functions? Assume that (1.11) does not hold for some multi-index $\gamma^{(0)}$ alone, where $|\gamma^{(0)}| \leq s$, so that

$$\sum_{\substack{|q|=0 \\ q\leq\gamma^{(0)}}}^{s-1} \frac{(1-i)^{\gamma^{(0)}-q}}{\left(\gamma^{(0)}-q\right)!}\,\omega_q(t+i) \not\equiv \frac{t^{\gamma^{(0)}}}{\gamma^{(0)}!} . \tag{3}$$

Let the function to be approximated be some $u \in C_0^{(s)}(\bar{\Omega})$. From (1.9) it follows that the approximation is valid if $u^{(\gamma_0)}(x) \equiv 0$.

Clearly, if the fundamental completeness conditions do not hold for multi-indices $\gamma^{(1)}, \gamma^{(2)}, \ldots, \gamma^{(K)}$, then the approximation is valid for functions for which

$$u^{\left(\gamma^{(l)}\right)}(x) \equiv 0 , \quad l = 1, 2, \ldots, K ,$$

are satisfied.

§4. On the Minimum Number of Primitive Functions[1]

We now pose the following question: does the system of coordinate functions (4.4) of Chapter I continue to be complete, if the number of primitive functions is decreased with respect to the condition $|q| \le s-1$? As we shall see below the answer is "no". Explicit formulations for it will be given later.

Consider a system of M linearly independent primitive functions $\omega_q(t)$ with $M < M_s$, where M_s is the number of multi-indices q such that $|q| \le s-1$ (see §4, Chapter I). For the set M , we enumerate the indices so that they contain the multi-index $\underline{0} = (0, 0, \ldots, 0)$ and do not contain the multi-index $\tilde{q} = (s-1, 0, \ldots, 0)$. The set of multi-indices q , corresponding to the given subset of primitive functions will be denoted by Q . We assume that the primitive functions satisfy conditions 1). and 2). of 4, Chapter I, and a condition analogous to 3). of 4, Chapter I:

$$\omega_q^{(\alpha)}(\underline{1}) = \delta_{\alpha q} , \quad \alpha, q \in Q . \tag{2}$$

We construct the coordinate system (4.4) of Chapter I. We prove that this system is not complete in the following sense: with respect to the topology $W_p^{(s)}(\Omega)$, not every function $u \in C_0^{(s)}(\overline{\Omega})$ can be approximated by the functions

$$u^h(x) = \sum_{q\in Q} \sum_{j\in \hat{J}^h} h^q u^{(q)}\big((j+\underline{1})h\big)\omega_q\left(\frac{x}{h} - j\right) . \tag{3}$$

Repeating the discussion of §1 and §3, we come to the conclusion: in order to ensure that every function $u \in C_0^{(s)}(\overline{\Omega})$ can be approximated, with respect to the topology of $W_p^{(s)}(\Omega)$, by the functions (3), it is necessary and sufficient that

$$\sum_{\substack{q\in Q\\ q\le\gamma}} \sum_{i\in I} \frac{(\underline{1}-i)^{\gamma-q}}{(\gamma-q)!}\,\omega_q^{(\alpha)}(t+i) = \delta_{\alpha\gamma} ,$$

$$\forall\alpha \text{ s.t. } |\alpha| = s , \quad \forall\gamma \text{ s.t. } |\gamma| \le s , \quad \forall t , \quad \underline{0} \le t \le \underline{1} . \tag{4}$$

Integrating (4), we obtain

$$\sum_{\substack{q\in Q\\ q\le\gamma}} \sum_{i\in I} \frac{(\underline{1}-i)^{\gamma-q}}{(\gamma-q)!}\,\omega_q(t+i) = \begin{cases} \varphi_\gamma(t) , & |\gamma| < s , \\ \dfrac{t^\gamma}{\gamma!} + \varphi_\gamma(t) , & |\gamma| = s , \end{cases} \tag{5}$$

where the $\varphi_\gamma(t)$ are polynomials of degree $s - 1$. We take the identity (5) in which $\gamma = \tilde{q} = (s-1, 0, \ldots, 0)$ and apply to it the operator D^β , where β is a

[1] See Mikhlin [14].

multi-index such that $|\beta| \leq s-1$. In the resulting identity we put $t = 0$:

$$\sum_{\substack{q\in Q\\ q\leq\tilde{q}}} \sum_{i\in I} \frac{(1-i)^{\tilde{q}-q}}{(\tilde{q}-q)!} \omega_q^{(\beta)}(i) = \varphi_{\tilde{q}}^{(\beta)}(\underline{0}) . \tag{6}$$

The identity (6) contains no contribution from the term with $i = \underline{1}$. In fact, we have $q \leq \tilde{q}$, so all the differences $\tilde{q}_l - q_l$, $l = 1, 2, \ldots, m$, are non-negative, the first of them being positive. Indeed, if $q_1 - q_1 = s - 1 - q_1 = 0$, then, since $|q| \leq s-1$, it follows that $q_2 = q_3 = \ldots = q_m = 0$ and $q = \tilde{q}$, which is impossible, because $\tilde{q} \notin Q$. Hence,

$$\underline{0}^{\tilde{q}-q} = \prod_{l=1}^{m} 0^{\tilde{q}_l - q_l} = 0 ,$$

and so the identity (6) does not contain a term with $i = \underline{1}$.

Any vector $i \in I$, $i \neq \underline{1}$, has at least one zero component, hence such a vector lies in one of the coordinate planes and it follows from (4.2), Chapter I, that $\omega_q^{(\beta)}(i) = 0$, for all $q \in Q$. Now (6) gives $\varphi_{\tilde{q}}^{(\beta)}(i) = 0$, for all β , $|\beta| \leq s-1$, and hence $\varphi_{\tilde{q}}(t) \equiv 0$. We now have

$$\sum_{\substack{q\in Q\\ q\leq\tilde{q}}} \frac{(1-i)^{\tilde{q}-q}}{(\tilde{q}-q)!} \omega_q(t+i) = 0 , \quad \forall t , \quad \underline{0} \leq t \leq \underline{1} . \tag{7}$$

We put $t = \underline{1}$ in (7). If $i \neq \underline{0}$, then at least one component of the vector $\underline{1} + i$ is equal to 2 . By (4.2), Chapter I, we have $\omega_q(1+i) = 0$. If $i = \underline{0}$, then we obtain using (2)

$$\omega_q(1) = \begin{cases} 1 , & q = \underline{0} , \\ 0 , & q \neq \underline{0} , \end{cases}$$

and the left hand side of (7) has the value $1/\tilde{q}!$: the contradiction which completes the proof.

§5. The Necessity of the Fundamental Completeness Conditions[1]

Let $\{\omega_q(t)\}$, $|q| \leq s-1$, denote the set of primitive functions. We have two possibilities:

(1) there are constants $b_{qi\gamma}$ such that

[1] See Mikhlin [18].

$$\sum_{\substack{|q|=0\\ q\le\gamma}}^{s-1} \sum_{i\in I} b_{qi\gamma}\omega_q(t+i) = \frac{t^\gamma}{\gamma!} + \varphi_\gamma(t) \ , \quad \forall\gamma \ , \quad |\gamma| \le s \ , \quad \forall t \ , \quad \underline{0} \le t \le \underline{1} \tag{1}$$

where the $\varphi_\gamma(t)$ denote polynomials of degree $|\gamma| - 1$;

(2) there are no such constants.

We prove that, when (1) holds, the primitive functions satisfy the fundamental completeness conditions. Initially, we show that constants $C_{qi\gamma}$ exist such that

$$\sum_{\substack{|q|=0\\ q\le\gamma}}^{s-1} \sum_{i\in I} C_{qi\gamma}\omega_q(t+i) = \frac{t^\gamma}{\gamma!} \ , \quad \forall\gamma \ , \quad |\gamma| \le s \ , \quad \forall t \ , \quad \underline{0} \le t \le \underline{1} \ . \tag{2}$$

This is certainly the situation when $\gamma = 0$, because we then have $\varphi_\gamma(t) \equiv 0$ and can set $C_{qi\underline{0}} = b_{qi\underline{0}}$. Now let us assume that we have proved the result for multi-indices γ such that $|\gamma| \le \bar{s}$, $0 \le \bar{s} < s-1$. Let $|\gamma| = \bar{s}+1$. Then the polynomial $\varphi_\gamma(t)$ can be expressed in the form $(A_\beta = \text{const.})$

$$\varphi_\gamma(t) = \sum_{|\beta|=0}^{\bar{s}} A_\beta \frac{t^\beta}{\beta!} = \sum_{|\beta|=0}^{\bar{s}} A_\beta \sum_{\substack{|q|=0\\ q\le\beta}}^{s-1} \sum_{i\in I} C_{qi\beta}\omega_q(t+i) = \sum_{|q|=0}^{\bar{s}} \sum_{i\in I} \omega_q(t+i) \sum_{\substack{|\beta|=0\\ \beta\le q}} C_{qi\beta}A_\beta$$

so the relations (2) hold if we put

$$C_{qi\gamma} = b_{qi\gamma} - \sum_{\substack{|\beta|=0\\ \beta\le q}} C_{qi\beta}A_\beta \ , \quad |\gamma| = \bar{s} + 1 \ .$$

Let β be a multi-index such that $|\beta| \le s-1$. We apply to both sides of (2) the operator D^β and then set $t = \underline{1} - i_1$, $i_1 \in I$. We then have

$$\sum_{\substack{|q|=0\\ q\le\gamma}}^{s-1} \sum_{i\in I} C_{qi\gamma}\omega_q^{(\beta)}(\underline{1}-i_1+i) = \frac{(\underline{1}-i_1)^{\gamma-\beta}}{(\gamma-\beta)!} \ . \tag{3}$$

If $i \ne i_1$, then at least one component of the vector $i - i_1$ is either 1 or -1 , and the corresponding component of the vector $1 - i_1 + i$ is either 2 or 0 . Consequently, $\omega_q^\beta(\underline{1}-i_1+i) = 0$, and the sum in (3) contains only terms with $i = i_1$. But it follows that $\omega_q^{(\beta)}(1-i_1+i) = \omega_q^{(\beta)}(1) = \delta_{\beta q}$, and the sum in (3) reduces to the single term $C_{\beta i_1\gamma}$, and hence

$$C_{\beta i_1 \gamma} = \frac{(1-i_1)^{\gamma-\beta}}{(\gamma-\beta)!}\ .$$

It only remains to interchange β and i_1 with q and i. Then (2) coincides with the fundamental completeness conditions.

We now turn to the second case, when the conditions (1) fail to hold for any values of the constants $b_{qi\gamma}$. Then, at least for one pair of multi-indices α and γ such that $|\alpha| = |\gamma| \le s$, the following inequality

$$C_0 \overset{\text{def}}{=\!=\!=} \inf_{C_{qi}} \int_Q \left| \delta_{\alpha\gamma} - \sum_{\substack{|q|=0 \\ q\le\gamma}}^{s-1} \sum_{i\in I} C_{qi}\omega_q^{(\alpha)}(t+i)\right|^p dt > 0 \tag{4}$$

holds, where here and below Q denotes the cube $\underline{0} \le t \le \underline{1}$.

Consider a bounded region Ω in $\mathbb{R}_m$ and assume that the primitive system $\{\omega_q(t)\}$ generates the coordinate system $\varphi_{qj}^{(h)}(x) = \omega_q\left(\frac{x}{h} - j\right)$, which is complete in $W_p^{(s)}(\Omega)$. We assume that $Q \subset \Omega$, since, otherwise, we can change the scale and redefine the origin. Consider a function $u_\gamma \in W_p^{(s)}(\Omega)$ such that $u_\gamma(x) = x^\gamma/\gamma!$, when $x \in Q$. Because the system $\left\{\varphi_{qj}^{(h)}\right\}$ is complete, there exist constants $a_{qj\gamma}^{(h)}$ such that

$$\left\| u_\gamma - u_\gamma^h \right\|_{W_p^{(s)}(Q)} \xrightarrow[h\to 0]{} 0 \tag{5}$$

where

$$u_\gamma^h(x) = \sum_{|q|=0}^{s-1} \sum_{j\in J^h} a_{qj\gamma}^{(h)}\omega_q\left(\frac{x}{h} - j\right)\ . \tag{6}$$

Let $1/2h$ be an integer. Then Q can be subdivided into smaller cubes. Let j_0 be the label of the lower vertex of one such cube, which we denote by Q_{j_0}. Consider the integral

$$S_{j_0} = \int_{Q_{j_0}} \left| \delta_{\alpha\gamma} - \sum_{|q|=0}^{s-1} \sum_{j\in J^h} h^{-|\alpha|} a_{qj\gamma}^{(h)} \omega\left(\frac{x}{h} - j\right)\right|^p dx\ ,$$

where α is the multi-index associated with (4). As usual, we transform S_{j_0} to

$$S_{j_0} = h^m \int_Q \left| \delta_{\alpha\gamma} - \sum_{|q|=0}^{s-1} \sum_{i\in I} h^{-|\alpha|} a_{q,j_0-i,\gamma}^{(h)} \omega_q^{(\alpha)}(t+i)\right|^p dt\ .$$

From (4) we obtain $S_{j_0} \geq h^m C_0$ and

$$\left\| u_\gamma - u_\gamma^h \right\|^p_{W_p^{(s)}(\Omega)} \geq C_0$$

which contradicts (5). The necessity of the fundamental completeness conditions is proved.

§6. One-Dimensional Primitive Systems

With $m = 1$, the fundamental completeness conditions (1.11) can be simplified. To begin with, the vectors t, q, γ, i become scalars with $0 \leq q \leq s-1$, $0 \leq \gamma \leq s$, $i = 0, 1$. Consider the fundamental completeness conditions (1.11) for the case when $\gamma = s$. Since $q \leq s-1$, then $\gamma - q = s - q > 0$ and all the terms with $i = 1$ vanish. The only remaining term corresponds to $i = 0$. As a result, we obtain

$$\sum_{q=0}^{s-1} \frac{\omega_q(t)}{(s-q)!} = \frac{t^s}{s!}, \quad 0 \leq t \leq 1 . \tag{1}$$

Now take $\gamma < s$. We must sum q from 0 to γ. One of the differences $\gamma - q$ equals zero while the others are positive, so if $i = 1$, then all the terms vanish except for the one with $q = \gamma$; that is, $\omega_\gamma(t+1)$. The conditions (1.11) thereby take the form

$$\omega_\gamma(t+1) + \sum_{q=0}^{\gamma} \frac{\omega_q(t)}{(\gamma-q)!} = \frac{t^\gamma}{\gamma!}, \quad 0 \leq \gamma \leq s-1 , \quad 0 \leq t \leq 1 . \tag{2}$$

Equations (1) and (2) yield the one-dimensional fundamental completeness conditions of degree $s - 1$. We recall that the primitive functions must also satisfy the conditions

$$\begin{aligned} &\omega_q^{(\alpha)}(1) = \delta_{\alpha q} , \\ &\omega_q^{(\alpha)}(0) = \omega_q^{(\alpha)}(2) = 0 , \quad 0 \leq \alpha, q \leq s-1 , \end{aligned} \tag{3}$$

which follow from conditions 2). and 3). of §4, Chapter I.

Our aim now is to describe the construction of one-dimensional primitive systems of arbitrary degree. It is sufficient to define the primitive functions on the interval $0 < t < 2$. Outside this interval they equal zero. This follows from the general definition of §4, Chapter I.

It is convenient to interpret the fundamental completeness conditions (1) and (2) as a system of $s + 1$ equations in the $2s$ unknowns $\omega_q(t)$ and $\psi_q(t) = \omega_q(t+1)$,

$q = 0, 1, \ldots, s-1$. These unknowns must satisfy the conditions defined by (3),

$$\omega_q^{(\alpha)}(0) = 0 \quad , \quad \omega_q^{(\alpha)}(1) = \delta_{\alpha q} \ , \tag{4_1}$$

$$\psi_q^{(\alpha)}(0) = \delta_{\alpha q}, \quad \psi_q^{(\alpha)}(1) = 0 \ , \quad 0 \le \alpha, \ q \le s-1 \ . \tag{4_2}$$

If $s = 1$, then the number of equations equals the number of unknowns, and the only solutions are $\omega_0(t) = t$, $\psi_0(t) = 1 - t$, $0 \le t \le 1$, which yields the single primitive function

$$\omega_0(t) = \begin{cases} t \ , & 0 \le t \le 1 \ , \\ 2-t \ , & 1 < t \le 2 \ , \\ 0 \ , & t \notin [0, 2] \ . \end{cases} \tag{5}$$

It is not difficult to verify that this function satisfies the conditions of (3) as well as the conditions of §4, Chapter I.

If $s > 1$, we can, for example, take the functions $\omega_0(t), \omega_1(t), \ldots, \omega_{s-2}(t)$, $0 \le t \le 1$, to be arbitrarily. They only have to be contained in $W_p^{(s)}(0, 1)$ and to satisfy (4_1) with $0 \le \alpha \le s-1$. The function $\omega_{s-1}(t)$ is then defined by (1), with the functions $\omega_q(t+1)$, $0 \le q \le s-1$, defined by (2). We now prove that such functions also satisfy conditions (4_1), (4_2).

We have from (1) that

$$\omega_{s-1}(t) = \frac{t^s}{s!} - \sum_{q=0}^{s-2} \frac{\omega_q(t)}{(s-q)!} \ .$$

Differentiating and putting $t = 0$ and $t = 1$, we obtain

$$\omega_{s-1}^{(\alpha)}(0) = 0 \ , \quad 0 \le \alpha \le s-1 \ ;$$

$$\omega_{s-1}^{(\alpha)}(1) = \begin{cases} 1 \ , & \alpha = s - 1 \ , \\ 0 \ , & 0 \le \alpha \le s-1 \ , \end{cases}$$

which yields (4_1). Then from (2) we have

$$\psi_\gamma(t) = \frac{t^\gamma}{\gamma!} - \sum_{q=0}^{\gamma} \frac{\omega_q(t)}{(\gamma-q)!} \ .$$

By this and (4_1) it evidently follows that

$$\psi_\gamma^{(\alpha)}(0) = \delta_{\alpha\gamma} \ , \quad \psi_\gamma^{(\alpha)}(1) = 0 \ , \quad 0 \le \alpha \ , \quad \gamma \le s-1 \ .$$

It is also obvious that $\omega_q \in W_p^{(s)}(0,\, 2)$.

We can set, for example,

$$\omega_q(t) = t^s P_q(t) \;, \quad 0 \le t \le 1 \;, \quad 0 \le q \le s-2 \;, \tag{6}$$

where $P_q(t)$ is a polynomial of degree $s - 1$ such that $\omega_q^{(\alpha)}(1) = \delta_{\alpha q}$, $0 \quad \alpha \quad s-1$. Obviously, in this situation,

$$\omega_{s-1}(t) = t^s P_s(t) \;, \quad \omega_q(t+1) = (1-t)^s \tilde{P}_q(t) \;, \quad 0 \le t \le 1 \;, \tag{7}$$

where $P_s(t)$ and $\tilde{P}_q(t)$ are also polynomials of the degree $s - 1$.

The polynomials $P_q(t)$, $0 \le q \le s-1$, can be constructed by solving some appropriate algebraic system. We express these polynomials in the form

$$P_q(t) = \sum_{k=0}^{s-1} \frac{a_{qk}}{(2s-k-1)!}\, t^{s-k-1} \;.$$

So, if $0 \le t \le 1$, then

$$\omega_q(t) = \sum_{k=0}^{s-1} \frac{a_{qk}}{(2s-k-1)!}\, t^{2s-k-1}$$

and the second condition in (4_1) yields the system

$$\sum_{k=0}^{s-1} \frac{a_{qk}}{(2s-k-\alpha-1)!} = \delta_{\alpha q} \;, \quad \alpha = 0,\, 1,\, \ldots,\, s-1 \;. \tag{8}$$

The determinant of (8) is non-zero. Otherwise, the corresponding homogeneous system would have a non-trivial solution. Then there would exist a polynomial $t^s P_q(t)$ of degree $2s - 1$, which is not identically equal to zero and has two multiple roots $t = 0$ and $t = 1$ with multiplicity s .

We note that the inversion of (8) generates all the polynomials $P_q(t)$, $0 \le q \le s-1$, and hence all the primitive functions $\omega_q(t)$, $0 \le q \le s-1$, on the interval $0 \le t \le 1$. In particular, there is no necessity to use (1) in order to find $\omega_{s-1}(t)$. Once the functions $\omega_q(t)$, $0 \le t \le 1$, have been constructed, the functions $\omega_q(t+1)$, $0 \le t \le 1$, can be constructed using (2).

We now show that primitive functions of the above form (6) and (7) can be defined independently of the inversion of (8) and without (2). We introduce the substitution $t = 1 + t$ and write

$$\omega_q(1+t) = \begin{cases} \sigma_q(t) \;, & -1 \le t \le 0 \;, \\ \psi_q(t) \;, & 0 \le t \le 1 \;. \end{cases}$$

The function $\sigma_q(t)$ is a polynomial of degree $2s-1$ satisfying the conditions

$$\sigma_q^{(\alpha)}(-1) = 0 \ , \quad \sigma_q^{(\alpha)}(0) = \delta_{\alpha q} \ , \quad 0 \le \alpha \ , \quad q \le s-1 \ , \tag{9}$$

which define it uniquely. Hence

$$\sigma_q(t) = (1+t)^s R_q(t) \tag{10}$$

where R_q is a polynomial of degree $s-1$. In particular, we may write

$$\sigma_0(t) = (1+t)^s \left[a_0^{(s)} - a_1^{(s)} t + a_2^{(s)} t^2 - \dots + (-1)^{s-1} a_{s-1}^{(s)} t^{s-1} \right] . \tag{11}$$

The second condition in (9) yields the recurrence relation

$$a_0^{(s)} = 1 \ , \quad \sum_{\nu=0}^{k} (-1)^{k-\nu} \binom{s}{k-\nu} a_\nu^{(s)} = 0 \ , \quad k > 0 \ , \tag{12}$$

which defines the function $\sigma_0(t)$. We now prove that

$$\sigma_q(t) = \frac{1}{q!} (1+t)^s t^q \left[a_0^{(s)} - a_1^{(s)} t + a_2^{(s)} t^2 - \dots + (-1)^{s-q-1} a_{s-q-1} t^{s-q-1} \right] . \tag{13}$$

Clearly, it is true when $q = 0$. We now assume that it holds for some $q \le s-2$ and prove that it holds for $q+1$. It is therefore sufficient to prove that the polynomial

$$\sigma_{q+1}(t) = \frac{1}{(q+1)!} (1+t)^s t^{q+1} \left[a_0 - a_1^{(s)} t + a_2^{(s)} t^2 - \dots + (-1)^{s-q-2} a_{s-q-2} t^{s-q-2} \right] \tag{14}$$

satisfies (9) with q replaced by $q+1$. The first set of conditions of (9) are obvious, so it remains to verify that

$$\sigma_{q+1}^{(\alpha)}(0) = \delta_{\alpha,q+1} \ , \quad 0 \le \alpha \le s-1 \ . \tag{15}$$

Equation (14) can be rewritten as

$$\sigma_{q+1}(t) = \frac{1}{q+1} t\sigma_q(t) - \frac{(-1)^{s-q-1}}{(q+1)!} a_{s-q-1}^{(s)} t^s (1+t)^s \ .$$

Hence

$$\sigma_{q+1}^{(\alpha)}(0) = \frac{1}{q+1} \left[t\sigma_q^{(\alpha)}(t) + \alpha\sigma_q^{(\alpha-1)}(t) \right]_{t=0} = \begin{cases} 0 \ , & \alpha \ne q+1 \ , \\ 1 \ , & \alpha = q+1 \ , \end{cases}$$

which proves (13).

We now examine the functions $\psi_q(t) = \omega_q(t+1)$, $0 \le t \le 1$. These are polynomials of degree $2s-1$ satisfying the identities

$$\psi_q^{(\alpha)}(0) = \delta_{\alpha q} \ , \quad \psi_q^{(\alpha)}(1) = 0 \ , \quad 0 \le \alpha \ , \quad q \le s-1 \ . \tag{16}$$

They guarantee that these polynomials are uniquely defined. In the same way, we can prove that

$$\psi_q(t) = \frac{1}{q!}(1-t)^s t^q \left[a_0^{(s)} + a_1^{(s)} t + a_2^{(s)} t^2 + \ldots + a_{s-q-1}^{(s)} t^{s-q-1}\right] , \tag{17}$$

where the coefficients $a_0^{(s)}$, $a_1^{(s)}$, ..., $a_{s-1}^{(s)}$ are defined by (12).

Below, in §6 of Chapter III, we mention an important property of primitive functions which have been constructed using (11)-(13) and (17).

In Table 1, some primitive systems are presented which have been constructed, for certain values of s, using (11)-(13) and (17).

In conclusion, we derive a formula which defines the solution of the recurrence (12). We determine the coefficients $b_k^{(s)}$ which satisfy the identity

$$\sum_{k=0}^{\infty} b_k^{(s)} Z^{(k)} \sum_{m=0}^{\infty} (-1)^m \binom{s}{m} Z^m = 1 .$$

Multiplying these two series and equating coefficients of the powers of Z, it follows that the coefficients $b_k^{(s)}$ satisfy (12) and $b_k^{(s)} = a_k^{(s)}$. On the other hand, we have

$$\sum_{m=0}^{\infty} (-1)^m \binom{s}{m} Z^m = (1-Z)^s .$$

Hence, it follows that

$$\sum_{k=0}^{\infty} a_k^{(s)} Z^k = (1-Z)^{-s} = 1 + \sum_{k=1}^{\infty} \frac{s(s+1)\ldots(s+k-1)}{k!} Z^k ,$$

which yields the required solution of (12):

$$a_0^{(s)} = 1 , \quad k > 0 , \quad a_k^{(s)} = \frac{s(s+1)\ldots(s+k-1)}{k!} . \tag{18}$$

For the evaluation of coefficients $a_k^{(s)}$, it is more convenient to use the recurrence relations which follows from (18):

$$a_0^{(s)} = 1 , \quad a_k^{(s)} = a_{k-1}^{(s)} + a_k^{(s-1)} , \quad 1 \le k \le s-2 ;$$
$$a_{s-1}^{(s)} = 2a_{s-2}^{(s)} . \tag{19}$$

§7. Primitive Systems of Higher Dimensions with Zero Degree

If the degree is zero $(s = 1)$, then the primitive system only contains one function $\omega_0(t)$. In this section, we denote it by $\omega(t) = \omega(t_1, t_2, \ldots, t_m)$.

TABLE 1

Primitive Functions

s	q	$-1 \leq t \leq 0$	$0 \leq t \leq 1$
1	0	$1 + t$	$1 - t$
2	0	$(1+t)^2(1-2t)$	$(1-t)^2(1+2t)$
	1	$(1+t)^2$	$(1-t)^2$
3	0	$(1+t)^3(1-3t+6t^2)$	$(1-t^3)(1+3t+6t^2)$
	1	$(1+t)^3 t(1-3t)$	$(1-t)^3 t(1+3t)$
	2	$\frac{1}{2}(1+t)^3 t^2$	$\frac{1}{2}(1-t)^3 t^2$
4	0	$(1+t)^4(1-4t+10t^2-20t^3)$	$(1-t)^4(1+4t+10t^2+20t^3)$
	1	$(1+t)^4 t(1-4t+10t^2)$	$(1-t)^4 t(1+4t+10t^2)$
	2	$\frac{1}{2}(1+t)^4 t^2(1-4t)$	$\frac{1}{2}(1-t)^4 t^2(1+4t)$
	3	$\frac{1}{6}(1+t)^4 t^3$	$-\frac{1}{6}(1-t)^4 t^3$
5	0	$(1+t)^5(1-5t+15t^2-35t^3+70t^4)$	$(1-t)^5(1+5t+15t^2+35t^3+70t^4)$
	1	$(1+t)^5 t(1-5t+15t^2-35t^3)$	$(1-t)^5 t(1+5t+15t^2+35t^3)$
	2	$\frac{1}{2}(1+t)^5 t^2(1-5t+15t^2)$	$\frac{1}{2}(1-t)^5 t^2(1+5t+15t^2)$
	3	$\frac{1}{6}(1+t)^5 t^3(1-5t)$	$\frac{1}{6}(1-t)^5 t^3(1+5t)$
	4	$\frac{1}{24}(1+t)^5 t^4$	$\frac{1}{24}(1-t)^5 t^4$
6	0	$(1+t)^6(1-6t+21t^2-56t^3+126t^4-252t^5)$	$(1-t)^6(1+6t+21t^2+56t^3+126t^4+252t^5)$
	1	$(1+t)^6 t(1-6t+21t^2-56t^3+126t^4)$	$(1-t)^6 t(1+6t+21t^2+56t^3+126t^4)$
	2	$\frac{1}{2}(1+t)^6 t^2(1-6t+21t^2-56t^3)$	$\frac{1}{2}(1-t)^6 t^2(1+6t+21t^2+56t^3)$
	3	$\frac{1}{6}(1+t)^6 t^3(1-6t+21t^2)$	$\frac{1}{6}(1-t)^6 t^3(1+6t+21t^2)$
	4	$\frac{1}{24}(1+t)^6 t^4(1-6t)$	$\frac{1}{24}(1-t)^6 t^4(1+6t)$
	5	$\frac{1}{120}(1+t)^6 t^5$	$-\frac{1}{120}(1-t)^6 t^5$

We examine in detail the case when $m = 2$. In this situation, the cube $\underline{0} \le t \le \underline{1}$ is the square with vertices (0, 0), (0, 1), (1, 0), (1, 1) . The vectors i can be identified with these vertices. The set of all such vectors yields the set I .

Consider (1.11). Since, for $m = 2$, $|\gamma| \le 1$, the multi-index γ has only three values: $\gamma = (0, 0)$, $\gamma = (0, 1)$, $\gamma = (1, 0)$. The multi-index q has the single value $\underline{0} = (0, 0)$, so $\gamma - q = \gamma$ and $(\gamma-q)! = \gamma!$ which equals unity for the mentioned values of γ .

We now construct a table of values $(1-i)^\gamma = (1-i_1)^{\gamma_1}(1-i_2)^{\gamma_2}$, where i_1, i_2 and γ_1, γ_2 are components of the vector i and the multi-index γ , respectively.

Since our aim is to consider the case $s = 2$ in the next section, we present in Table 2 the values of $(\underline{1}-i)^\gamma$ for $|\gamma| \le 2$.

TABLE 2

The values $(\underline{1}-i)^\gamma$

i \ γ	(0, 0)	(0, 1)	(1, 0)	(0, 2)	(1, 1)	(2, 0)
(0, 0)	1	1	1	1	1	1
(0, 1)	1	0	1	0	0	1
(1, 0)	1	1	0	1	0	0
(1, 1)	1	0	0	0	0	0

Corresponding to the three values of γ , we obtain from (1.11) with the help of Table 2 three equations which relate the functions $\omega(t_1, t_2)$, $\omega(t_1+1, t_2)$, $\omega(t_1, t_2+1)$ and $\omega(t_1+1, t_2+1)$ on the square $Q_0\{(t_1, t_2) : 0 \le t_1, t_2 \le 1\}$ (see Figure 13):

$$\omega(t_1, t_2) + \omega(t_1, t_2+1) = t_1 , \tag{1}$$

$$\omega(t_1, t_2) + \omega(t_1+1, t_2) = t_2 , \tag{2}$$

$$\omega(t_1, t_2) + \omega(t_1+1, t_2) + \omega(t_1, t_2+1) + \omega(t_1+1, t_2+1) = 1 . \tag{3}$$

We now see that if the function $\omega(t_1, t_2)$ is defined on the square Q_0 , then its values on the squares Q_1, Q_2 and Q_3 are defined by (see Figure 13)

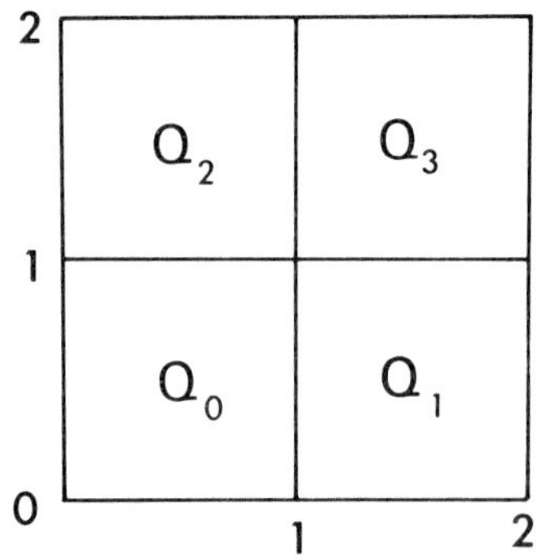

Fig. 13

$$\left.\begin{aligned}
\psi_1(t_1, t_2) &\overset{\text{def}}{=\!=} \omega(t_1+1, t_2) = t_2 - \omega(t_1, t_2) , \\
\psi_2(t_1, t_2) &\overset{\text{def}}{=\!=} \omega(t_1, t_2+1) = t_1 - \omega(t_1, t_2) , \\
\psi_3(t_1, t_2) &\overset{\text{def}}{=\!=} \omega(t_1+1, t_2+1) = 1 - t_1 - t_2 + \omega(t_1, t_2) .
\end{aligned}\right\} \tag{4}$$

The function $\omega(t_1, t_2)$, defined on Q_0 , must satisfy the boundary conditions (4.2) of Chapter I and conditions 3). of §4, Chapter I, which now take the form

$$\omega(0, t_2) = \omega(t_1, 0) = 0 \tag{5}$$

and

$$\omega(1, 1) = 1 . \tag{6}$$

The functions (4) must satisfy analogous conditions, and ω must be continuous across the common sides of the squares Q_k . The latter yields additional conditions to be imposed on $\omega(t_1, t_2)$.

For the functions ψ_k , conditions 3). of §4, Chapter I, take the form

$$\psi_1(0, 1) = \psi_2(1, 0) = \psi_3(0, 0) = 1 .$$

Equations (4)-(6) show that these conditions are satisfied. We now consider the boundary conditions. The function $\psi_1(t_1, t_2) = \omega(t_1+1, t_2)$ must vanish when $t_1 = 1$ and $t_2 = 0$, so setting $t_1 = 1$ and $t_2 = 0$ in the first equation in (4), we obtain the identities

$$t_2 - \omega(1, t_2) = 0 \ , \quad \omega(t_1, 0) = 0 \ .$$

The second coincides with one of the equations in (5), the first gives a new condition which ω must satisfy

$$\omega(1, t_2) = t_2 \ . \tag{7}$$

In the same way, we find from the second equation in (4) that

$$\omega(t_1, 1) = t_1 \ . \tag{8}$$

It is not difficult to see that the third equation of (4) does not involve new conditions. It can be easily verified that if $\omega(t_1, t_2)$ is defined and continuous on the square Q_0 and satisfies conditions (5), (7) and (8) on its boundary, and if we consider this function to be defined on Q_1, Q_2 and Q_3 by (4), then such a function is continuous on the square $\underline{0} \le t \le \underline{2}$.

Finally, for $m = 2$ and $s = 1$, we can construct a primitive function in the following way. We consider a continuous function $\omega(t_1, t_2)$ on the square Q_0 , which has generalised first derivatives which are summable for a given power p , $1 \le p \le \infty$. This function must satisfy conditions (5), (7) and (8) and is defined on the squares Q_1, Q_2 and Q_3 using (4).

We can construct a primitive function on Q_0 as follows: we join the point (0, 0) to any point on the boundary of Q_0 by a straight line. The function ω is known at the ends of this line, and between them we define it by linear interpolation. This yields the "pyramid" function, considered in §2, Chapter I. It is not difficult to construct the following representation for this primitive function

$$\begin{aligned}
&0 \le t_1 \ , \ t_2 \le 1 \ ; \ \omega(t_1, t_2) = \begin{cases} t_2 \ , & t_2 \le t_1 \ , \\ t_1 \ , & t_1 \le t_2 \ ; \end{cases} \\
&1 \le t_1 \le 2 \ , \ 0 \le t_2 \le 1 \ ; \ \omega(t_1, t_2) = \begin{cases} 0 \ , & t_2 \le t_1 - 1 \ , \\ 1 - t_1 + t_2 \ , & t_2 \ge t_1 - 1 \ ; \end{cases} \\
&0 \le t_1 \le 1 \ , \ 1 \le t_2 \le 2 \ ; \ \omega(t_1, t_2) = \begin{cases} 1 + t_1 - t_2 \ , & t_2 \le t_1 + 1 \ , \\ 0 \ , & t_2 \ge t_1 + 1 \ ; \end{cases} \\
&1 \le t_1 \ , \ t_2 \le 2 \ ; \ \omega(t_1, t_2) = \begin{cases} 2 - t_1 \ , & t_2 \le t_1 \ , \\ 2 - t_2 \ , & t_2 \ge t_1 \ . \end{cases}
\end{aligned} \tag{9}$$

Simple primitive functions of the following form can be constructed

$$0 \le t_1 \,,\ t_2 \le 1 \;;\ \omega(t_1, t_2) = t_1 t_2 \,. \tag{10}$$

From (4), it follows that

$$\begin{aligned} 1 \le t_1 \le 2 \,,\ 0 \le t_2 \le 1 \;;\ \omega(t_1, t_2) &= (2-t_1)t_2 \;; \\ 0 \le t_1 \le 1 \,,\ 1 \le t_2 \le 2 \;;\ \omega(t_1, t_2) &= t_1(2-t_2) \;; \\ 1 \le t_1 \,,\ t_2 \le 2 \;;\ \omega(t_1, t_2) &= (2-t_1)(2-t_2) \,. \end{aligned} \tag{10_1}$$

Obviously, (10) and (10_1) define the primitive function which satisfies

$$\omega(t_1, t_2) = \omega_0(t_1)\omega_0(t_2) \tag{11}$$

where ω_0 denotes the one-dimensional primitive function of zero degree defined by (C.5). In an analogous way, we can construct primitive functions of zero degree and arbitrary dimension m : it is sufficient to write

$$\omega(t) = \omega_{\underline{0}}(t) = \prod_{k=1}^{m} \omega_0(t_k) \,. \tag{12}$$

§8. Primitive Systems with $m = s = 2$

With $m = s = 2$, the conditions $|q| \le s-1$ and $|\gamma| \le s$ yield three possible values for the multi-index q : (0, 0), (1, 0), (0, 1) , and six possible values for the multi-index γ : (2, 0), (1, 1), (0, 2), (1, 0), (0, 1), (0, 0) . We thereby obtain three primitive functions $\omega_{00}(t_1, t_2)$, $\omega_{10}(t_1, t_2)$ and $\omega_{01}(t_1, t_2)$ which must satisfy the following boundary conditions

$$\begin{aligned} \omega_q(0, t_2) = \omega_q(2, t_2) = \omega_q(t_1, 0) = \omega_q(t_1, 2) = 0 \,, \\ \omega_{q/1}(0, t_2) = \omega_{q/1}(2, t_2) = \omega_{q/2}(t_1, 0) = \omega_{q/2}(t_1, 2) = 0 \,, \end{aligned} \quad . \tag{1}$$

where the subscript notation "$/i$" , $i = 1, 2$, implies differentiation with respect to the ith coordinate direction. In addition, these functions must satisfy the six equations of (1.11), corresponding to the six values of γ .

In this section, we assume that $0 \le t_1 \,,\ t_2 \le 1$.

Using Table 2 of §7, we rewrite the equations of (1.11) as

$$\omega_{00}(t_1, t_2) + \omega_{00}(t_1, t_2+1) + 2\omega_{10}(t_1, t_2) + 2\omega_{10}(t_1, t_2+1) = t_1^2 \,, \tag{2}$$

$$\omega_{00}(t_1, t_2) + \omega_{10}(t_1, t_2) + \omega_{10}(t_1+1, t_2) + \omega_{01}(t_1, t_2) + \omega_{01}(t_1, t_2+1) = t_1 t_2 \,, \tag{3}$$

$$\omega_{00}(t_1, t_2) + \omega_{00}(t_1+1, t_2) + 2\omega_{01}(t_1, t_2) + 2\omega_{01}(t_1+1, t_2) = t_2^2 \,, \tag{4}$$

$$\omega_{00}(t_1, t_2) + \omega_{00}(t_1, t_2+1) + \omega_{10}(t_1, t_2) + \\ + \omega_{10}(t_1+1, t_2) + \omega_{10}(t_1, t_2+1) + \omega_{10}(t_1+1, t_2+1) = t_1 , \quad (5)$$

$$\omega_{00}(t_1, t_2) + \omega_{00}(t_1+1, t_2) + \omega_{01}(t_1, t_2) + \\ + \omega_{01}(t_1+1, t_2) + \omega_{01}(t_1, t_2+1) + \omega_{01}(t_1+1, t_2+1) = t_2 , \quad (6)$$

$$\omega_{00}(t_1, t_2) + \omega_{00}(t_1+1, t_2) + \omega_{00}(t_1, t_2+1) + \omega_{00}(t_1+1, t_2+1) = 1 . \quad (7)$$

We now construct the matrix defined by (2)-(7) with the unknowns arranged in the following order:

(1) $\omega_{00}(t_1, t_2)$; (2) $\omega_{00}(t_1+1, t_2)$; (3) $\omega_{00}(t_1, t_2+1)$;

(4) $\omega_{00}(t_1+1, t_2+1)$; (5) $\omega_{10}(t_1, t_2)$; (6) $\omega_{10}(t_1+1, t_2)$;

(7) $\omega_{10}(t_1, t_2+1)$; (8) $\omega_{10}(t_1+1, t_2+1)$; (9) $\omega_{01}(t_1, t_2)$;

(10) $\omega_{01}(t_1+1, t_2)$; (11) $\omega_{01}(t_1, t_2+1)$; (12) $\omega_{01}(t_1+1, t_2+1)$.

It becomes

	1	2	3	4	5	6	7	8	9	10	11	12	
1	1	0	1	0	2	0	2	0	0	0	0	0	t_1^2
2	1	0	0	0	1	1	0	0	1	0	1	0	t_1t_2
3	1	1	0	0	0	0	0	0	2	2	0	0	t_2^2
4	1	0	1	0	1	1	1	1	0	0	0	0	t_1
5	1	1	0	0	0	0	0	0	1	1	1	1	t_2
6	1	1	1	1	0	0	0	0	0	0	0	0	1

(8)

Columns 4, 8 and 12 contain unit in the 6th, 4th and 5th rows, respectively, with the remaining elements zeros. Consequently, the functions

$$\omega_{00}(t_1+1, t_2+1) , \quad \omega_{10}(t_1+1, t_2+1) , \quad \omega_{01}(t_1+1, t_2+1) ,$$

can be determined from equations (7), (5) and (6), once the other functions are known. We note that these three functions are independent of the other equations of the system (2)-(7). The first three rows of the matrix (8) contain a non-zero minor consisting of the elements of its first three columns; in fact

$$\begin{vmatrix} 1 & 0 & 1 \\ 1 & 0 & 0 \\ 1 & 1 & 0 \end{vmatrix} = 1 .$$

Hence, it follows that we can define the functions

$$\omega_{00}(t_1,\ t_2)\ ,\quad \omega_{00}(t_1+1,\ t_2)\ ,\quad \omega_{00}(t_1,\ t_2+1)\ ,$$

using the equations (2)-(4).

We now come to the conclusion that: if the functions ω_{10} and ω_{01} are defined arbitrarily on the squares Q_0, Q_1, Q_2 (see Figure 13), then they are uniquely defined on Q_3 by (2). In addition, the values of ω_{00} on all squares Q_0, Q_1, Q_2, Q_3 can be determined. In fact, we have

$$\omega_{00}(t_1,\ t_2) = \omega_{10}(t_1,\ t_2) - \omega_{10}(t_1+1,\ t_2) - \omega_{01}(t_1,\ t_2) - \omega_{01}(t_1,\ t_2+1) + t_1t_2\ ; \tag{9}$$

$$\omega_{00}(t_1+1,\ t_2) = \omega_{10}(t_1,\ t_2) + \omega_{10}(t_1+1,\ t_2) - \omega_{01}(t_1,\ t_2) - 2\omega_{01}(t_1+1,\ t_2) + \omega_{01}(t_1,\ t_2+1) + t_2^2 - t_1t_2\ ; \tag{10}$$

$$\omega_{00}(t_1,\ t_2+1) = -\omega_{10}(t_1,\ t_2) + \omega_{10}(t_1+1,\ t_2) - 2\omega_{10}(t_1,\ t_2+1) + \omega_{01}(t_1,\ t_2) + \omega_{01}(t_1,\ t_2+1) + t_1^2 - t_1t_2\ ; \tag{11}$$

$$\omega_{00}(t_1+1,\ t_2+1) = \omega_{10}(t_1,\ t_2) - \omega_{10}(t_1+1,\ t_2) + 2\omega_{10}(t_1,\ t_2+1) + \omega_{01}(t_1,\ t_2) + 2\omega_{01}(t_1+1,\ t_2) - \omega_{01}(t_1,\ t_2+1) + 1 - t_1^2 + t_1t_2 - t_2^2\ ; \tag{12}$$

$$\omega_{10}(t_1+1,\ t_2+1) = \omega_{10}(t_1,\ t_2) - \omega_{10}(t_1+1,\ t_2) + \omega_{10}(t_1,\ t_2+1) - t_1^2 + t_1\ ; \tag{13}$$

$$\omega_{01}(t_1+1,\ t_2+1) = \omega_{01}(t_1,\ t_2) + \omega_{01}(t_1+1,\ t_2) - \omega_{01}(t_1,\ t_2+1) - t_2^2 + t_2\ . \tag{14}$$

Since the functions ω_q must satisfy the boundary conditions (1), it follows from (9)-(14) that the boundary conditions for ω_{10} and ω_{01} on the lines $t_1 = 1$ and $t_2 = 1$, $0 \le t_1$, $t_2 \le 1$, are given by

$$\left.\begin{aligned} \omega_{10}(1,\ t_2) &= 0\ , & \omega_{10}(1,\ t_2+1) &= 0\ , \\ \omega_{10/1}(1,\ t_2) &= t_2\ , & \omega_{10/1}(1,\ t_2+1) &= 1 - t_2\ ; \end{aligned}\right\} \tag{15}$$

$$\left.\begin{aligned} \omega_{01}(t_1,\ 1) &= 0\ , & \omega_{01}(t_1+1,\ 1) &= 0\ , \\ \omega_{01/2}(t_1,\ 1) &= t_1\ , & \omega_{01/2}(t_1+1,\ 1) &= 1 - t_1\ ; \end{aligned}\right\} \tag{16}$$

$$\left.\begin{aligned} \omega_{10}(t_1,\ 1) - \omega_{10}(t_1+1,\ 1) &= t_1^2 - t_1\ , \\ \omega_{10/2}(t_1,\ 1) - \omega_{10/2}(t_1-1,\ 1) &= 0\ ; \end{aligned}\right\} \tag{17}$$

$$\left.\begin{aligned} \omega_{01}(1,\, t_2) - \omega_{01}(1,\, t_2+1) &= t_2^2 - t_2\,, \\ \omega_{01/1}(1,\, t_2) - \omega_{01/1}(1,\, t_2+1) &= 0\,. \end{aligned}\right\} \qquad (18)$$

If (9)-(14) and the boundary conditions (15)-(18) are satisfied, then the functions ω_q and their first derivatives are condinuous on the lines $t_1 = 1$ and $t_2 = 1$ and satisfy the conditions 3)., §4 of Chapter I, at the point $t_1 = t_2 = 1$.

Now, in order to construct the primitive functions, we must construct the functions $\omega_{10}(t_1,\, t_2)$, $\omega_{10}(t_1+1,\, t_2)$, $\omega_{10}(t_1,\, t_2+1)$, $\omega_{01}(t_1,\, t_2)$, $\omega_{01}(t_1+1,\, t_2)$ and $\omega_{01}(t_1,\, t_2+1)$ which satisfy the boundary conditions (1) and (15)-(18). It is sufficient to construct the three functions with subscripts "10" and then, by symmetry, obtain the functions with subscripts "01" by interchanging the arguments t_1 and t_2.

The construction of the functions ω_{10} can be accomplished in the following way. The values of $\omega_{10}(t_1,\, 1)$ and $\omega_{10/2}(t_1,\, 1)$ are chosen arbitrarily such that they satisfy the first condition of (15). For example, we can set $\omega_{10}(t_1,\, 1) = t_1^2(t_1-1)$ and $\omega_{10/2}(t_1,\, 1) = 0$. Then the values of all three functions and their normal derivatives will be known on the sides of the square Q_0. In fact, they are:

The function $\omega_{10}(t_1,\, t_2)$

$$\left.\begin{aligned} \omega_{10}(0,\, t_2) = \omega_{10}(1,\, t_2) &= 0\,, \\ \omega_{10}(t_1,\, 0) = 0\,, \quad \omega_{10}(t_1,\, 1) &= t_1^2(t_1-1)\,, \\ \omega_{10/1}(0,\, t_2) = 0\,, \quad \omega_{10/1}(1,\, t_2) &= t_2\,, \\ \omega_{10/2}(t_1,\, 0) = \omega_{10/2}(t_1,\, 1) &= 0\,; \end{aligned}\right\} \qquad (19)$$

The function $\omega_{10}(t_1+1,\, t_2)$

$$\left.\begin{aligned} \omega_{10}(1,\, t_2) = \omega_{10}(2,\, t_2) &= 0\,, \\ \omega_{10}(t_1+1,\, 0) = 0\,, \quad \omega_{10}(t_1+1,\, 1) &= t_1(1-t_1)^2\,, \\ \omega_{10/1}(1,\, t_2) = t_2\,, \quad \omega_{10/1}(2,\, t_2) &= 0\,, \\ \omega_{10/2}(t_1+1,\, 0) = \omega_{10/2}(t_1+1,\, 1) &= 0\,; \end{aligned}\right\} \qquad (20)$$

The function $\omega_{10}(t_1, t_2+1)$

$$\left.\begin{aligned} &\omega_{10}(0, t_2+1) = \omega_{10}(1, t_2+1) = 0 , \\ &\omega_{10}(t_1, 1) = t_1^2(t_1-1) , \quad \omega_{10}(t_1, 2) = 0 , \\ &\omega_{10/1}(0, t_2+1) = 0 , \quad \omega_{10/1}(1, t_2+1) = 1 - t_2 , \\ &\omega_{10/2}(t_1, 1) = \omega_{10/2}(t_1, 2) = 0 . \end{aligned}\right\} \tag{21}$$

We next construct the function $\omega_{10}(t_1, t_2)$ which satisfies the conditions (19). The others can be constructed in analogous ways.

We shall determine $\omega_{10}(t_1, t_2)$ in the form

$$\omega_{10}(t_1, t_2) = t_1^2 t_2^2 \Phi(t_1, t_2) . \tag{22}$$

Since the conditions of (19) involving a zero argument are automatically satisfied, the remaining conditions yield the following four restrictions on the form of $\Phi(t_1, t_2)$:

$$\begin{aligned} &\Phi(1, t_2) = 0 , \quad \Phi_{/1}(1, t_2) = \frac{1}{t_2} , \\ &\Phi(t_1, 1) = t_1 - 1 , \quad \Phi_{/2}(t_1, 1) = 2(1-t_1) . \end{aligned} \tag{23}$$

It thereby follows from (23) that it is impossible to construct $\omega_{10}(t_1, t_2)$ as a polynomial.

The first two conditions of (23) are satisfied by putting

$$\Phi(t_1, t_2) = \frac{t_1-1}{1-t_1+t_2} f(t_1, t_2) \tag{24}$$

with

$$f(1, t_2) = 1 . \tag{25}$$

The last two conditions of (23) yield

$$f(t_1, 1) = 2 - t_1 , \quad f_{/2}(t_1, 1) = 2t_1 - 3 . \tag{26}$$

Combining conditions (25) and (26), we obtain

$$f(t_1, t_2) = \frac{(1-t_1)(1-t_2)(3-2t_1)}{2-t_1-t_2} + 2 - t_1 . \tag{27}$$

In this way, the final form of $\omega_{10}(t_1, t_2)$ on Q is found to be

$$\omega_{10}(t_1, t_2) = \frac{t_1^2 t_2^2 (t_1-1)}{1-t_1+t_2}\left[2 - t_1 + \frac{(1-t_1)(1-t_2)(3-2t_1)}{2-t_1-t_2}\right] . \quad (28)$$

Introducing polar coordinates first with center at $(1, 0)$, and then with center at $(1, 1)$, we can verify that (28) is continuously differentiable on $\overline{Q_0}$, and its second derivatives are bounded and continuous in Q_0 .

In Mikhlin [8, 9], the equations corresponding to (27) and (28) are incorrect.

§9. Product Primitive Systems

As we saw in the preceeding sections, the construction of primitive systems satisfying the fundamental completeness conditions can be performed rather simply, and the primitive functions have rather simple structure if the dimension of the system is equal to unity or if its degree is zero. When $m = s = 2$, the primitive system turns out to be rather cumbersome.

Using special one-dimensional coordinate functions, Aubin [1] developed the method of constructing multidimensional coordinate functions as the products of one-dimensional ones. Unfortunately, the basic inequality, which implies the completeness of the constructed system, is only derived for functions of some non-dense set; so Aubin's resulting proof of completeness would appear to be defective.

In Mikhlin [17], the product of one-dimensional functions is considered for primitive functions satisfying the fundamental completeness conditions (6.1) and (6.2). We now reproduce the essence of the argument developed there.

Let t be a real variable, and let $\omega_0(t), \omega_1(t), \ldots, \omega_{s-1}(t)$ be a one-dimensional primitive system of degree $s - 1$ satisfying the fundamental completeness conditions (6.1), (6.2). If $s = 1$, then we set

$$\tilde{\omega}_{\underline{0}}(t) = \prod_{k=1}^{m} \omega_0(t_k) .$$

This case is considered in §7. The primitive function $\tilde{\omega}_0(t)$ satisfies the fundamental completeness conditions (1.11), and hence, the corresponding coordinate system is complete in $W_p^{(s)}$. Below, we assume that $s > 1$.

Taking the product of one-dimensional primitive functions, we can construct primitive functions of dimension m of the form

$$\tilde{\omega}_q(t) = \prod_{k=1}^{m} \omega_{q_k}(t_k) , \quad t = (t_1, t_2, \ldots, t_m) ,$$
$$q = (q_1, q_2, \ldots, q_m) , \quad \underline{0} \le q \le s\text{-}\underline{1} . \quad (1)$$

It is clear that these functions satisfy

$$\tilde{\omega}_q \in C^{(s-1)}(R_m) \cap W_p^{(s)}(R_m) \ ; \quad \operatorname{supp} \tilde{\omega}_q \subset \{t : \underline{0} \le t \le \underline{2}\} \ ;$$

$$\tilde{\omega}_q^{(\alpha)}(1) = \delta_{\alpha q} \ , \quad \underline{0} \le \alpha \ , \quad q \le s\text{-}\underline{1} \ .$$

From (1), we can construct the coordinate functions

$$\varphi_{qj}^{(h)}(x) = \tilde{\omega}_q\left(\frac{x}{h} - j\right) . \tag{2}$$

To each sufficiently smooth function $u(x)$, there corresponds a function

$$u^h(x) = \sum_{q=\underline{0}}^{s-1} \sum_{j\in y^h} h^q u^{(q)}\big((j+1)h\big)\tilde{\omega}_q\left(\frac{x}{h} - j\right) , \tag{3}$$

which can be considered as an approximation to $u(x)$.

If we only require that the $W_p^{(s)}(\Omega)$ norm of the difference should be small, then we can eliminate the items for which $|q| > s$ in (3), since, with respect to the $W_p^{(s)}(\Omega)$ norm, they tend to zero as $h \to 0$ and do not influence the character of the approximations generated by (3). Hence, it follows that the products (1) only need to be used for those multi-indices q which satisfy the inequality

$$|q| \le s \ . \tag{4}$$

So, instead of (3), we examine the approximations

$$u^h(x) = \sum_{|q|=0}^{s} \sum_{j\in J^h} h^q u^{(q)}\big((j+1)h\big)\tilde{\omega}_q\left(\frac{x}{h} - j\right) . \tag{5}$$

In the same way as in §1 and §3 we can prove that the coordinate system (2) is complete in $W_p^{(s)}(\Omega)$, and the function (5) approximates a sufficiently smooth function $u(x)$, if conditions, analogous to (1.11), hold:

$$\sum_{q\le\gamma} \sum_{i\in I} \frac{(1-i)^{\gamma-q}}{(\gamma-q)!} \tilde{\omega}_q(t+i) = \frac{t^\gamma}{\gamma!} \ , \quad |\gamma| \le s \ , \quad \underline{0} \le t \le \underline{1} \ . \tag{6}$$

We now prove that (6) in fact holds. We can write (6.1) and (6.2) for the one-dimensional primitive functions $\omega_0(t), \omega_1(t), \ldots, \omega_{s-1}(t)$ in the form

$$\sum_{\substack{q_k=0\\ q_k\le\gamma_k}}^{s-1} \sum_{i_k=0,1} \frac{\left(1-i_k\right)^{\gamma_k-q_k}}{(\gamma_k-q_k)!} \omega_{q_k}\left(t_k\right) = \frac{t_k^{\gamma_k}}{\gamma_k!} \ , \tag{7}$$

where $k = 1, 2, \ldots, m$, $0 \le \gamma_k \le s$.

We take an arbitrary multi-index $\gamma = (\gamma_1, \gamma_2, \ldots, \gamma_m)$, with $|\gamma| \le s$, write (7) for each of the values $\gamma_1, \gamma_2, \ldots, \gamma_m$, multiply them and put $q = (q_1, q_2, \ldots, q_m)$. As a result we obtain that the functions (1) satisfy the conditions

$$\sum_{q \le \gamma} \sum_{i \in I} \frac{(1-i)^{\gamma-q}}{(\gamma-q)!} \tilde{\omega}_q(t+i) = \frac{t^\gamma}{\gamma!} , \quad \underline{0} \le \gamma \le \underline{s} . \tag{8}$$

The conditions (6) follows from (8), if the last are considered only for the values of γ for which $|\gamma| \le s$. So the completeness of the system (2) is proved.

The primitive system

$$\tilde{\omega}_q(t) = \prod_{k=1}^{m} \omega_{q_k}(t_k) , \quad t = (t_1, t_2, \ldots, t_m) ,$$
$$q = (q_1, q_2, \ldots, q_m) , \quad |q| \le s , \tag{9}$$

will be referred to as a *product system*.

CHAPTER III

ORDER OF APPROXIMATION

In the previous chapter, it was shown that functions from Sobolev spaces can be approximated by linear combinations of coordinate functions which are obtained from primitive functions satisfying fundamental convergence conditions. In this chapter, we study the dependence of the order of approximation on the smoothness of the functions being approximated.

§1. Order of Approximation using the Uniform Norm

1°. Let $u \in C^{(s+1)}(\overline{\Omega})$. We assume Ω is such that $u(x)$ can be extended to R_m and is still contained in the same space. As noted in §2, Chapter II, the support of the extended function can be regarded as compact.

We construct a parallelepiped Π, with sides parallel to the coordinate axes, which contains the support of the extended function. The extension of $u(x)$ on Π will be denoted by $u(x)$. Independently of Π, we construct a rectangular grid with side h and denote the parallelepipeds, which are the union of the larger cubes lying in Π, by $\hat{\Pi}^h$. If h is sufficiently small, then $\operatorname{supp} u \in \hat{\Pi}^h$ and the norms of $u(x)$ in $C^{(s+1)}(\Pi)$ and $C^{(s+1)}(\hat{\Pi}^h)$ will coincide.

We denote by $\hat{J}^h_\Pi$ the labels of the lower vertices of the larger mesh cubes which define $\hat{\Pi}^h$, and denote by $u^h(x)$ the interpolatory function

$$u^h(x) = \sum_{|q|=0}^{s-1} \sum_{j \in \hat{J}^h_\Pi} h^q u^{(q)}\big((j+1)h\big)\omega_q\left(\frac{x}{h} - j\right) . \tag{1}$$

The goal of this section is the construction of a proof of the following assertion:

For any integer $\bar{s}$, $0 \le \bar{s} \le s$, *the inequality*

$$\|u-u^h\|_{C^{(\bar{s})}(\Pi)} \le C\|u\|_{C^{(s+1)}(\Pi)} h^{s-\bar{s}+1} \tag{2}$$

holds, where C *depends only on* m, s *and* $\bar{s}$.

Consider the difference

$$D^\alpha u(x) - D^\alpha u^h(x) \ , \quad |\alpha| = s_0 \le \bar{s} \ , \quad x \in \hat{\Pi}^h \ .$$

Assume that x is located in the smaller cube with lower vertex $j_0 h$. Then $x = x_0 + th$, where $x_0 = j_0 h$, $\underline{0} \le t \le \underline{1}$. As shown in §1, Chapter II, the non-zero terms in the sum (1) are those for which $j = j_0 - i$, $i \in I$, and the sum (1) can therefore be reduced to the form

$$u^h(x) = \sum_{|q|=0}^{s-1} \sum_{i \in I} h^q u^{(q)}\big(x_0+(1-i)h\big)\omega_q(t+i) \ .$$

Hence

$$D^\alpha u(x) - D^\alpha u^h(x) = u^{(\alpha)}\big(x_0+th\big) - \sum_{|q|=0}^{s-1} \sum_{i \in I} h^{|q|-s_0} u^{(q)}\big(x_0+(1-i)h\big)\omega_q^{(\alpha)}(t+i) \ . \tag{3}$$

With respect to powers of h , we develop the Taylor series expansion of $u^{(\alpha)}\big(x_0+th\big)$ up to terms in h^{s-s_0} , and of $u^{(q)}\big(x_0+(1-i)h\big)$ up to terms in $h^{s-|q|}$. For functions of many variables, the Taylor series takes the form

$$w(x+\xi) = \sum_{|\beta|=0}^{h} \frac{\xi^\beta}{\beta!} w^{(\beta)}(x) + \rho_n \ , \tag{4}$$

where the remainder

$$\rho_n = (n+1) \int_0^1 \sum_{|\beta|=n+1} \frac{\xi^\beta}{\beta!} w^{(\beta)}(x+z\xi)(1-z)^n dz \ . \tag{5}$$

Since $|\alpha| = s_0$, we obtain

$$u^{(\alpha)}\big(x_0+th\big) = \sum_{|\beta|=0}^{s-s_0} \frac{h^\beta t^\beta}{\beta!} u^{(\alpha+\beta)}\big(x_0\big) +$$

$$+ \big(s-s_0+1\big)h^{s-s_0+1} \int_0^1 \sum_{|\beta|=s-s_0+1} \frac{t^\beta}{\beta!} u^{(\alpha+\beta)}\big(x_0+tzh\big)(1-z)^{s-s_0} dz \ ,$$

$$u^{(q)}\left(x_0+(\underline{1}-i)h\right) = \sum_{|\beta|=0}^{s-|q|} \frac{h^{\beta}(\underline{1}-i)^{\beta}}{\beta!} u^{(\beta+q)}\left(x_0\right) +$$

$$+ (s-|q|+1)h^{s-|q|+1} \int_0^1 \sum_{|\beta|=s-|q|+1} \frac{(\underline{1}-i)^{\beta}}{\beta!} u^{(\beta+q)}\left(x_0+(\underline{1}-i)hz\right)(1-z)^{s-|q|}dz \ . \qquad (6)$$

We substitute these values in (3), and note that the remainders have values which are dominated by

$$C \max_{|\mu|=s_0+1} \|u^{(\mu)}\|_{C(\Pi)} h^{s-s_0+1} \qquad (7)$$

where C depends on m, s_0 and s. We now show that the sum of all the other terms in (3) is equal to zero. The sum in question is

$$\sum_{|\beta|=0}^{s-s_0} \frac{h^{\beta}t^{\beta}}{\beta!} u^{(\alpha+\beta)}\left(x_0\right) - \sum_{|q|=0}^{s-1} \sum_{|\beta|=0}^{s-q} \frac{h^{|\beta+q|-s_0}(\underline{1}-i)^{\beta}}{\beta!} u^{(\beta+q)}\left(x_0\right)\omega_q^{(\alpha)}(t+i) \ . \qquad (8)$$

We eliminate β from (8) by putting $\alpha+\beta=\gamma$ in the first sum and $\beta+q=\gamma$ in the second. The expression (8) takes the form

$$\sum_{|\gamma|=s_0}^{s} h^{|\gamma|-s_0} u^{(\gamma)}\left(x_0\right) \frac{t^{\gamma-\alpha}}{(\gamma-\alpha)!} - \sum_{|q|=0}^{s-1} \sum_{|\gamma|=|q|}^{s} \sum_{i\in I} h^{|\gamma|-s_0} u^{(\gamma)}\left(x_0\right) \frac{(\underline{1}-i)^{\gamma-q}}{(\gamma-q)!} \omega_q^{(\alpha)}(t+i) =$$

$$= \sum_{|\gamma|=s_0}^{s} h^{|\gamma|-s_0} u^{(\gamma)}\left(x_0\right) \frac{t^{\gamma-\alpha}}{(\gamma-\alpha)!} - \sum_{|\gamma|=0}^{s} \sum_{\substack{|q|=0\\ q\le\gamma}}^{s-1} \sum_{i\in I} h^{\gamma-s_0} u^{(\gamma)}\left(x_0\right) \frac{(\underline{1}-i)^{\gamma-q}}{(\gamma-q)!} \omega_q^{(\alpha)}(t+i) =$$

$$= \sum_{|\gamma|=s_0}^{s} h^{|\gamma|-s_0} u^{(\gamma)}\left(x_0\right) \left[\frac{t^{\gamma-\alpha}}{(\gamma-\alpha)!} - \sum_{\substack{|q|=0\\ q\le\gamma}}^{s-1} \sum_{i\in I} \frac{(\underline{1}-i)^{\gamma-q}}{(\gamma-q)!} \omega_q^{(\alpha)}(t+i)\right] -$$

$$- \sum_{|\gamma|=0}^{s_0-1} h^{|\gamma|-s_0} u^{(\gamma)}\left(x_0\right) \sum_{\substack{|q|=0\\ q\le\gamma}}^{s-1} \sum_{i\in I} \frac{(\underline{1}-i)^{\gamma-q}}{(\gamma-q)!} \omega_q^{(\alpha)}(t+i) \ .$$

It is not difficult to verify that, by applying D^{α}, $|\alpha| = s_0$, to both sides of the conditions (1.11) of Chapter I, the last expression in the above equation equals zero. Now

$$|D^{\alpha}u(x)-D^{\alpha}u^h(x)| \le C \max_{|\mu|=s+1} \|u^{(\mu)}\|_{C(\Pi)} h^{s-s_0+1} \le$$

$$\le C\|u\|_{C^{(s+1)}(\Pi)} h^{s-s_0+1} \ , \quad |\alpha| = s_0 \ , \quad C = \text{const.}$$

Taking the sum over α, $|\alpha| \le \bar{s}$, we obtain (2).

We note the following corollary. Obviously we have

$$\|u-u^h\|_{C^{(\bar{s})}(\Omega)} \le \|u-u^h\|_{C^{(\bar{s})}(\Pi)} .$$

On the other hand, the Whitney and Hestenes extension theorem for smooth functions yields

$$\|u\|_{C^{(s+1)}(\Pi)} \le C_0\|u\|_{C^{(s+1)}(\bar{\Omega})} ,$$

where $C_0 = C_0(\Omega, \Pi, m, s)$. Setting $C_1 = CC_0$, we derive the inequality which defines the accuracy of the approximation in Ω :

$$\|u-u^h\|_{C^{(\bar{s})}(\bar{\Omega})} \le C_1\|u\|_{C^{(s+1)}(\bar{\Omega})} h^{s-\bar{s}+1} , \quad 0 \le \bar{s} \le s . \tag{9}$$

2°. Now let $u \in C^{(s,\lambda)}(\bar{\Omega})$, $0 < \lambda \le 1$. We recall that the norm in $C^{(s,\lambda)}(\bar{\Omega})$ is defined to be

$$\|u\|_{C^{(s,\lambda)}(\bar{\Omega})} = \sum_{|\alpha|=0}^{s} \max_{x\in\bar{\Omega}} |u^{(\alpha)}(x)| + \sum_{|\alpha|=s} \sup_{x,y\in\bar{\Omega}} \frac{|u^{(\alpha)}(x)-u^{(\alpha)}(y)|}{|x-y|^{\lambda}} .$$

As before, we assume that $u(x)$ can be extended to $\mathbb{R}_m$, so that the extended function is contained in the same space and its support is compact and lies inside the parallelpiped Π , described in 1°. The extended function is again denoted by u . We now prove that, in this situation, inequality (2) is changed in the following way:

$$\|u-u^h\|_{C^{(\bar{s})}(\Pi)} \le C\|u\|_{C^{(s,\lambda)}(\Pi)} h^{s-\bar{s}+\lambda} , \quad 0 \le \bar{s} \le s . \tag{10}$$

As a consequence, inequality (9) is changed in the same way.

The Taylor series (4) remains valid for $w \in C^{(n,\lambda)}(\bar{G})$, where $G \subset \mathbb{R}_n$ is an arbitrary bounded domain, but the remainder becomes

$$\rho_n = n \int_0^1 \sum_{|\beta|=n} \frac{\xi^\beta}{\beta!} \left[u^{(\beta)}(x_0+z\xi)-u^{(\beta)}(x_0)\right](1-z)^{n-1}dz . \tag{11}$$

In fact, for $w \in C^{(n,\lambda)}(\bar{G})$, (4) and (5) become, after changing n to $n-1$,

$$w(x+\xi) = \sum_{|\beta|=0}^{n-1} \frac{\xi^\beta}{\beta!} w^{(\beta)}(x) + n \int_0^1 \sum_{|\beta|=n} \frac{\xi^\beta}{\beta!} w^{(\beta)}(x+z\xi)(1-z)^{n-1}dz . \tag{12}$$

We rewrite the second integral as follows:

$$n \int_0^1 \sum_{|\beta|=n} \frac{\xi^\beta}{\beta!} w^{(\beta)}(x+z\xi)(1-z)^{n-1}dz =$$

$$= n \int_0^1 \sum_{|\beta|=n} \frac{\xi^\beta}{\beta!} [w^{(\beta)}(x+z\xi)-w^{(\beta)}(x)](1-z)^{n-1}dz + \sum_{|\beta|=n} \frac{\xi^\beta}{\beta!} w^{(\beta)}(x) .$$

Using the last term in (12), we find that $w(x+\xi)$ is the sum of the principal Taylor terms of (4) and the remainder (11). We note that this remainder is dominated by

$$C\|w\|_{C^{(n,\lambda)}(\bar{G})} |\xi|^{n+\lambda} . \tag{13}$$

The rest of the proof reduces to a repetition (with small changes) of the reasoning presented in 1°. Consider (3) and take the Taylor series expansion of its right hand side. The principal terms vanish because of the fundamental completeness condition, and from (13) the remainders yield (10).

3°. The extension to functions $u \in C^{(s+1)}(\bar{\Omega})$ and $C^{(s,\lambda)}(\bar{\Omega})$ can be accomplished only if appropriate conditions are imposed on the boundary $\partial\Omega$. If we perform the extension, then, proceeding in the same way as above, we can prove that if $u \in C^{(s,\lambda)}(\bar{\Omega})$, $0 < \lambda \le 1$, and K is an arbitrary compact set in Ω , then

$$\begin{gathered} \|u-u^h\|_{C^{(\bar{s})}(K)} \le C\|u\|_{C^{(s,\lambda)}(\bar{\Omega})} h^{s-\bar{s}+\lambda} , \\ C = C(\Omega,\ K,\ m,\ s,\ \bar{s},\ \lambda) . \end{gathered} \tag{14}$$

In the same way we can prove the following, more general, statements:

(a) If the function $u \in C^{(s)}(\bar{\Omega})$ be extended to $\mathbb{R}_m$ so that it belongs to the same space and u^* denotes the ex[illegible]d function, then

$$\|u-u^h\|_{C^{(\bar{s})}(\bar{\Omega})} \le Ch^{s-\bar{s}}\sigma_s(u^*,\ h) , \tag{15}$$

where $\sigma_s(u^*,\ h)$ is the modulus of continuity of the s-order derivatives of the extended function u^* .

(b) If $u \in C^{(s)}(\bar{\Omega})$ and K is a compact set in Ω , then

$$\|u-u^h\|_{C^{(\bar{s})}(K)} \le Ch^{s-\bar{s}}\sigma_s(u,\ h) , \tag{16}$$

where C can now depend on K , and $\sigma_s(u,\ h)$ is the modulus of continuity of the s-order derivatives of the function u in $\bar{\Omega}$.

§2. On the Averaging of Functions

Sobolev averaging, often used above, gives mean functions which tend slowly to the corresponding averaged function when the averaging radius tends to zero. The method of averaging proposed by Il'in [1] and developed by Golovkin[1] [1] does not

[1] Golovkin (1936-1969), who died prematurely, was a Leningrad mathematician who published many significant papers on functional analysis and partial differential equations.

suffer from this shortcoming. The new method represents a refinement of Sobolev's method. We now present Golovkin's results as they apply to the norms of $W_p^{(s)}$, since this is of central interest in the sequel. In this examination, we start from Golovkin's basic ideas, but use other arguments.

Let $\varphi(x)$ denote an infinitely often differentiable function in $\mathbb{R}_m$ and let its support be contained in the ball $|x| \le 1$. We assume that $\varphi(x)$ satisfies the following conditions

$$\int_{\mathbb{R}_m} x^\mu \varphi(x)\,dx = \begin{cases} 1, & \mu = 0, \\ 0, & 1 \le |\mu| \le s. \end{cases} \tag{1}$$

Such a function has been constructed by Golovkin [1]. We set

$$\tau_h(x) = h^{-m}\varphi\left(\frac{x}{h}\right), \tag{2}$$

and call the function $\tau_h(x)$ the *averaging kernel of order* s. We note some simple properties of this averaging kernel: $\operatorname{supp} \tau_h \subset \{x : |x| \le h\}$;

$$\int_{\mathbb{R}_m} \tau_h(x)\,dx = 1; \quad \int_{\mathbb{R}_m} |\tau_h(x)|\,dx = c < \infty;$$
$$\int_{\mathbb{R}_m} x^\mu \tau_h(x) = 0, \quad 1 \le |\mu| \le s. \tag{3}$$

Let $u \in L(\Omega)$, where Ω is now an arbitrary measurable set in $\mathbb{R}_m$. We extend $u(x)$ to $\mathbb{R}_m$ by taking it equal to zero outside Ω. We set

$$\bar{u}_h(x) = \int_{\mathbb{R}_m} u(y)\,\tau_h(y-x)\,dy \tag{4}$$

and call $\bar{u}_h(x)$ the mean function for $u(x)$. The following properties of mean functions are obvious:

$$\bar{u}_h(x) = \int_{r<h} u(y)\,\tau_h(y-x)\,dy = \int_{\Omega} u(y)\,\tau_h(y-x)\,dy =$$
$$= \int_{\Omega \cap (r<h)} u(y)\,\tau_h(y-x)\,dy, \quad r = |y-x|; \tag{5}$$

$\bar{u}_h \in C^\infty(\mathbb{R}_m)$; if $u \in L_p(\Omega)$, $1 \le p \le \infty$, then $\|\bar{u}_h\|_p \le c\|u\|_p$, where c is the same as in (3) and $\|\cdot\|_p$ is the norm in $L_p(\Omega)$.

THEOREM. *Let* $u \in W_p^{(s+1)}(\mathbb{R}_m)$, $1 \le p \le \infty$ *and assume that the support of* u *is compact in* $\mathbb{R}_m$. *Then*

$$\|u-\bar{u}_h\|_{p,\bar{s}} \le C\|u\|_{p,s+1} h^{s-\bar{s}+1} , \quad 0 \le \bar{s} \le s , \quad C = \text{const.} \tag{6}$$

It is sufficient to prove the theorem for $\bar{s} = 0$.

The derivatives $u^{(\alpha)}(x)$, $|\alpha| \le s + 1$, exist for almost all $x \in R_m$. The same derivatives are absolutely continuous along almost all rays coming from x , when $|\alpha| \le s$. Hence it follows that for almost all points $x, y \in R_m$ the Taylor series

$$u(y) = u(x) + \sum_{|\beta|=0}^{s} \frac{(y-x)^\beta}{\beta!} u^{(\beta)}(x) + \rho_s(x, y) \tag{7}$$

holds with remainder

$$\rho_s(x, y) = (s+1) \int_0^1 \sum_{|\beta|=s+1} \frac{(y-x)^\beta}{\beta!} u^{(\beta)}\big(x+z(y-x)\big)(1-z)^s dz . \tag{8}$$

Multiplying (7) by $\tau_h(y-x)$ and integrating, we obtain

$$\bar{u}_h(x) = u(x) + \int_{r<h} \rho_s(x, y)\tau_h(y-x)dy .$$

Hence, for $1 < p < \infty$, with $1/p + 1/p' = 1$,

$$|\bar{u}_h(x)-u(x)| \le \left\{\int_{r<h} |\rho_s(x, y)|^p |\tau_h(y-x)|dy\right\}^{1/p} \times$$
$$\times \left\{\int_{r<h} |\tau_h(y-x)|dy\right\}^{1/p'} \le C^{1/p'}\left\{\int_{r<h} |\rho_s(x, y)|^p |\tau_h(y-x)|dy\right\}^{1/p} .$$

The supports of $u(x)$ and $\bar{u}_h(x)$ are compact and hence contained in some ball $|x| < a$. Integrating over this ball, we obtain

$$\|\bar{u}_h-u\|_p^p \le C^{p/p'} \int_{|x|<a} dx \int_{r<h} |\rho_s(x, y)|^p \cdot |\tau_h(y-x)|dy .$$

Using $\rho_s(x, y)$ from (8), we obtain

$$\|\bar{u}_h-u\|_p^p \le C^{p-1}(s+1)^p \int_{|x|<a} dx \int_{r<h} \left|\int_0^1 \sum_{|\beta|=s+1} \frac{(y-x)^\beta}{\beta!} \times\right.$$
$$\left.\times\, u^{(\beta)}\big(x+z(y-x)\big)(1-z)^s dz\right|^p \cdot |\tau_h(y-x)|dy .$$

Putting $y = x + th$, it follows that

$$\|\bar{u}_h-u\|_p^p \le C^{p-1}(s+1)^p h^{p(s+1)+m} \int_{|x|<a} dx \int_{|t|<1} \left|\int_0^1 \sum_{|\beta|=s+1} \frac{t^\beta}{\beta!} \times\right.$$
$$\left.\times\, u^{(\beta)}(x+zht)(1-z)^s dz\right|^p \cdot |\tau_h(th)|dt .$$

We now eliminate x by setting $x + zht = \xi$. Here $|\xi| \le a + h$ and $u(\xi) \equiv 0$ when $|\xi| > a$, so it is sufficient to integrate over the ball $|\xi| < a$:

$$\|\bar{u}_h - u\|_p^p \le C^{p-1}(s+1)^p h^{p(s+1)} \times$$

$$\times \int_{|\xi|<a} d\xi \int_{|t|<1} \left| \int_0^1 \sum_{|\beta|=s+1} \frac{t^\beta}{\beta!} u^{(\beta)}(\xi)(1-z)^s dz \right|^p \cdot |\varphi(t)| dt$$

$$\le C_1^p h^{p(s+1)} \|u\|_{p,s+1}^p , \quad C_1 = \text{const.}$$

Extracting the pth root, we obtain (6) for $\bar{s} = 0$. The theorem can be proved for $p = 1$ and $p = \infty$ in analogous but simpler ways.

§3. The Order of Approximation for Sobolev Spaces

THEOREM. *Let* $u \in W_p^{(s+1)}(\Omega)$, $1 < p \le \infty$, *where* Ω *is a finite region in* $\mathbb{R}_m$, *and let* $u(x)$ *be extended to* $\mathbb{R}_m$ *so that* $u \in W_p^{(s+1)}(\mathbb{R}_m)$. *Then there exist coefficients* $a_{qj}^{(h)}$ *such that*

$$\|u - u^h\|_{p,\bar{s}} \le C\|u\|_{p,s+1} h^{s-\bar{s}+1} , \quad 0 \le \bar{s} \le s , \tag{1}$$

where

$$\|\cdot\|_{p,\sigma} = \|\cdot\|_{W_p^{(\sigma)}(\Omega)} ,$$

and

$$u^h(x) = \sum_{|q|=0}^{s-1} \sum_{j \in \bar{J}^h} a_{qj}^{(h)} \omega_q\left(\frac{x}{h} - j\right) .$$

For $p = 2$, this theorem was proved by Strang and Fix [2]. Their proof is essentially based on a Fourier transform analysis. The general case was studied by Demjanovic and Mikhlin [1] using rather elementary tools. Below we develop their analysis.

Consider the case $1 < p < \infty$. If $p = 1$ or $p = \infty$, the arguments are in general the same, but the technology is simpler.

Let the support of the extended function $u(x)$ be located in some open cube Q with sides parallel to the coordinate axes. We assume h to be less than the distance between points of the sets ∂Q and $\operatorname{supp} u$. We average $u(x)$, using the averaging kernel $\tau_h(x)$ of order s (see §2), and construct the mean function $\bar{u}_h(x)$. We construct the interpolatory function $v^h(x) = \bar{u}_h^h(x)$ for $\bar{u}_h(x)$ using (5.1) of Chapter 1. We take as $\bar{J}$ the labels of the lower vertices of the larger cubes in Q . The union of all these cubes is taken to be $Q^h \subset Q$. By (1.9) we have, for any domain $G \subset Q$, the inequality

$$\left\|\bar{u}_h - v^h\right\|_{C^{(\bar{s})}(\bar{G})} \le C_1 \|\bar{u}_h\|_{C^{(s+1)}(\bar{G})} h^{s-\bar{s}+1} . \tag{2}$$

We take as G the samller cube Q_{j_0} with label j_0.

Let $x \in Q_{j_0}$. Integrating by parts, we obtain

$$\bar{u}_h^{(\alpha)}(x) = \int_{r<h} u^{(\alpha)}(y)\tau_h(y-x)dy .$$

Hence, by Holder inequality,

$$\left|\bar{u}_h^{(\alpha)}(x)\right| \le \left\{\int_{r<h} |u^{(\alpha)}(y)|^p dy\right\}^{1/p} \left\{\int_{r<h} |\tau_h(y-x)|^{p'} dy\right\}^{1/p'} .$$

The second multiplier on the right hand side is dominated by $Ch^{-m/p}$, C = const., so

$$\left|\bar{u}_h^{(\alpha)}(x)\right|^p \le C^p h^{-m} \int_{r<h} |u^{(\alpha)}(y)|^p dy .$$

Summing over α, we obtain

$$\sum_{|\alpha|=0}^{\bar{s}} \left|\bar{u}_h^{(\alpha)}(x)\right|^p \le C^p h^{-m} \int_{r<h} \sum_{|\alpha|=0}^{\bar{s}} |u^{(\alpha)}(y)|^p dy .$$

It now follows from (2) that for all $x \in Q_{j_0}$

$$\sum_{|\alpha|=0}^{\bar{s}} \left|\bar{u}_h^{(\alpha)}(x) - v^{h(\alpha)}(x)\right|^p \le C_1^p h^{p(s-\bar{s}+1)-m} \int_{r<h} \sum_{|\beta|=0}^{s-1} |u^{(\beta)}(y)|^p dy . \tag{3}$$

We integrate the inequality (3) over Q_{j_0} and change the order of integration on the right hand side. As x ranges over the cube Q_{j_0}, y ranges over the domain B_{j_0}, contained in the cube $Q^*_{j_0}$. The length of the side of $Q^*_{j_0}$ equals zh and its lower vertex is $(j_0-1)h$ (see Figure 14). We therefore obtain the inequality

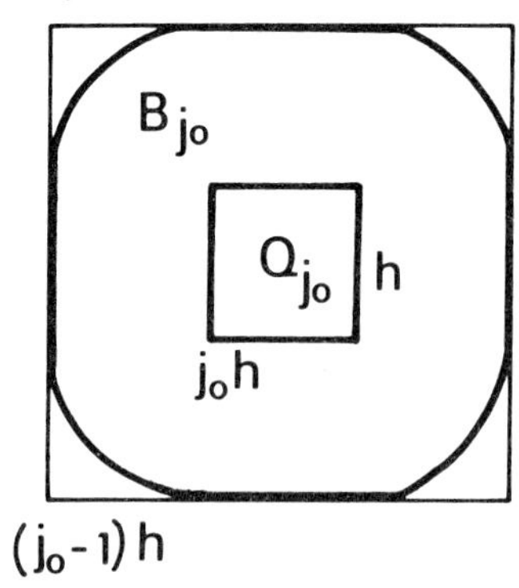

Fig. 14

$$\int_{Q_{j_0}} \sum_{|\alpha|=0}^{\bar{s}} \left|\bar{u}_h^{(\alpha)}(x)-v^{h(\alpha)}(x)\right|^p dx \le c^p h^{p(s-\bar{s}+1)} \int_{Q_{j_0}^*} \sum_{|\beta|=0}^{s+1} |u^{(\beta)}(y)|^p dy .$$

Summing over all j_0 such that $Q_{j_0} \subset Q^h$, we obtain the new inequality

$$\int_{Q^h} \sum_{|\alpha|=0}^{\bar{s}} \left|u_h^{(\alpha)}(x)-v^{h(\alpha)}(x)\right|^p dx \le 3^m c_1^p h^{s-\bar{s}+1} \int_{Q^h} \sum_{|\beta|=0}^{s+1} |u^{(\beta)}(y)|^p dy .$$

We drop the integral over $Q^h\backslash\Omega$ on the left handside. Using the definition of the extended function, the integral on the right hand side is dominated by some constant multiplied by $\|u\|_{p,s+1}^p$, and we thereby obtain the inequality

$$\left\|\bar{u}_h-v^h\right\|_{p,\bar{s}} \le C_2 h^{s-\bar{s}+1}\|u\|_{p,s+1} .$$

Hence, the inequality (1) follows from (2.5), if we set $u^h = v^h = \bar{u}_h^h$. Consequently

$$a_{qj}^{(h)} = h^q \bar{u}_h^{(q)}\left((j+\underline{1})h\right) .$$

NOTE. A sufficient, but not necessary, condition for $u(x)$ to be extended is $\partial\Omega \in C^{(0,1)}$. If, for example, $u(x)$ satisfies the boundary conditions

$$u^{(\alpha)}(x)|_{\partial\Omega} = 0 , \quad |\alpha| \le s , \tag{4}$$

then it is sufficient to define it to be zero outside Ω. The extension belongs to $W_p^{(s+1)}(\mathbb{R}_m)$, and its support is compact.

It is not difficult to obtain an estimate for $u(x) \in W_p^{(s)}(\Omega)$ when its extension has a modulus of continuity of order s. This is left as an exercise for the reader.

§4. Estimation of the Constants for the Simplest Case

If $m = \bar{s} = s = 1$ and if the norm in $W_p^l(a, b)$ is defined by

$$\|u\|_{p,l}^p = \int_a^b \sum_{k=1}^{l} |u^{(k)}(x)|^p dx ,$$

then we can take $c = 1$ in (3.1). We now prove this. Since $m = 1$, it is not necessary to take the average value of the function. We define the approximations by

$$u^h(x) = \sum_{j=-1}^{2n-1} u\left((j+1)h\right)\omega_0\left(\frac{x}{h} - j\right) , \tag{1}$$

where $\omega_0(t)$ is the function (6.5) of Chapter II. We consider the segment $[a, b]$ and assume $u \in W_p^{(1)}(a, b)$. The graph of $u^h(x)$ is the piecewise linear curves with vertices $(x_j, u(x_j))$, $x_j = a + jh$, $j = 0, 1, \ldots, 2n$, $h = (b-a)/2n$. Consider the integral

$$F = \int_a^b |u(x)-u^h(x)|^p dx = \sum_{k=0}^{2n-1} F_k ,$$

where

$$F_k = \int_{x_k}^{x_{k+1}} |u(x)-u^h(x)|^p dx .$$

The graph of (1) on (x_k, x_{k+1}) is the straight line joining the points $(x_k, u(x_k))$ and $(x_{k+1}, u(x_{k+1}))$. Hence

$$u^h(x) = u(x_k) + \frac{x-x_k}{h}\left[u(x_{k+1})-u(x_k)\right] , \quad x_k \le x \le x_{k+1} . \tag{2}$$

Consequently, on the same interval

$$u(x) - u(x_k) = \int_{x_k}^{x_{k+1}} K(x, t)u'(t)dt ,$$

$$u(x_{k+1}) - u(x_k) = \int_{x_k}^{x_{k+1}} u'(t)dt ,$$

$$K(x, t) = \begin{cases} 0 , & x < t , \\ 1 , & x \ge t , \end{cases}$$

and hence

$$F_k = \int_{x_k}^{x_{k+1}} \left| \int_{x_k}^{x_{k+1}} \left[K(x, t) - \frac{x-x_k}{h}\right] u'(t)dt \right|^p dx .$$

The value in square brackets is not greater than 1 and not less than -1 so, by the Hölder inequality,

$$\left| \int_{x_k}^{x_{k+1}} \left[K(x, t) - \frac{x-x_k}{h}\right] u'(t)dt \right|^p \le h^{p-1} \int_{x_k}^{x_{k+1}} |u'(t)|^p dt .$$

It follows then that

$$F_k \le h^p \int_{x_k}^{x_{k+1}} |u'(t)|^p dt \; .$$

Summing over k, we obtain

$$\int_a^b |u(x)-u^h(x)|^p dx \le h^p \int_a^b |u'(x)|^p dx \; . \tag{3}$$

We now assume that $u \in W_p^{(2)}(a,\, b)$. Consider the integral

$$E = \int_a^b |u'(x)-u^{h\prime}(x)|^p dx = \sum_{k=0}^{2n-1} E_k \; ,$$

$$E_k = \int_{x_k}^{x_{k+1}} |u'(x)-u^{h\prime}(x)|^p dx \; .$$

By (2) we have

$$E_k = \int_{x_k}^{x_{k+1}} \left| u'(x) - \frac{u(x_{k+1})-u(x_k)}{h} \right|^p dx \; .$$

Then, using Taylor series,

$$u(x_{k+1}) - u(x_k) = hu'(x_k) + \int_{x_k}^{x_{k+1}} (x_{k+1}-t)u''(t)dt \; .$$

At the same time

$$u'(x) - u'(x_k) = \int_{x_k}^{x_{k+1}} K(x,\, t)u''(t)dt \; .$$

Hence,

$$E_k = \int_{x_k}^{x_{k+1}} \left| \int_{x_k}^{x_{k+1}} \left[K(x,\, t) - \frac{x_{k+1}-t}{h} \right] u''(t)dt \right|^p dx \; .$$

As before the value in square brackets is not greater than 1 and not less than -1, so

$$E_k \le h^p \int_{x_k}^{x_{k+1}} |u''(x)|^p dx \; .$$

Summing, we obtain the new inequality

$$\int_a^b |u'(x)-u^{h\prime}(x)|^p dx \le h^p \int_a^b |u''(x)|^p dx \ . \tag{4}$$

We take the sum of (3) and (4)

$$\|u-u^h\|_{p,1}^p \le h^p \int_a^b [|u'(x)|^p+|u''(x)|^p]\,dx \ , \tag{5}$$

which yields

$$\|u-u^h\|_{p,1} \le h\|u\|_{p,2} \ , \tag{6}$$

and hence, the required result.

More accurate estimates of F_k and E_k show that C can be taken equal to 1/2 , when $p = 2$.

§5. Approximation Using Product Primitive Functions

In this section, we prove a result analogous to the Theorem of §3. Let $u \in W_p^{(s+1)}(\Omega)$ and assume that $u(x)$ has been extended to $\mathbb{R}_m$ so that $u \in W_p^{(s+1)}(\mathbb{R}_m)$. Let $\tilde{\omega}_q(t)$, $|q| \le s$, denote the product primitive functions described in §9 of Chapter 2. We set

$$u^h(x) = \sum_{|q|=0}^{s} \sum_{j\in J^h} a_{qj}^{(h)} \tilde{\omega}_q\left(\frac{x}{h} - j\right) , \quad a_{qj}^{(h)} = \text{const.} \tag{1}$$

We prove that the constants $a_{qj}^{(h)}$ can be chosen so that

$$\|u-u^h\|_{p,\bar{s}} \le Ch^{s-\bar{s}+1}\|u\|_{p,s+1} \ , \quad 0 \le \bar{s} \le s \ , \quad C = \text{const.} \tag{2}$$

We assume that $u(x)$ has been extended to $\mathbb{R}_m$ so that $u \in W_p^{(s+1)}(\mathbb{R}_m)$ and the support of $u(x)$ is compact. We take the Golovkin average of $u(x)$ (see §2). If $v_h(x) = \bar{u}_h(x)$ denotes the averaged function, then

$$\|u-v_h\|_{p,\bar{s}} \le C_1 h^{s-\bar{s}+1}\|u\|_{p,s+1} \ , \quad C_1 = \text{const.}, \tag{3}$$

and it suffices to prove that (2) holds for v_h .

Let Q be an open cube with sides parallel to the coordinate axes and such that supp $u \subset Q$. If the grid size h is sufficiently small, then supp $v_h \subset Q$. Consider firstly the following case. Let $v \in C^{(s+1)}(\mathbb{R}_m)$ and supp $v \subset Q$. We set

$$v^h(x) = \sum_{|q|=0}^{s} \sum_{j\in\hat{J}^h} h^q v^{(q)}\left((j+\underline{1})h\right)\tilde{\omega}_q\left(\frac{x}{h} - j\right) , \tag{4}$$

where $\hat{J}^h$ denotes the labels of the lower vertices of the larger cubes in Q. We denote by $\hat{Q}^h$ the union of these cubes. Clearly $\hat{Q}^h \subset Q$. We denote by $\hat{J}_0^h$ the labels of the lower vertices of the small cubes in $\hat{Q}^h$, and let Q_{j_0} denote the small cube with vertex $j_0 h$ $\left(j_0 \in \hat{J}_0^h\right)$. Consider the difference

$$\Delta = D^\alpha v(x) - D^\alpha v^h(x) , \quad x \in Q_{j_0} , \quad |\alpha| = s_0 \le \bar{s} .$$

We set $x = x_0 + th$, $x_0 = j_0 h$, with $\underline{0} \le t \le \underline{1}$. The difference Δ can be rewritten in the form (compare with §1 of Chapter II)

$$v^{(\alpha)}\left(x_0+th\right) - \sum_{|q|=0}^{s} \sum_{i\in I} h^{|q|-s_0} v^{(q)}\left(x_0+(\underline{1}-i)h\right)\tilde{\omega}_q(t+i) . \tag{5}$$

As before, we take the Taylor series expansion for $v^{(\alpha)}\left(x_0+th\right)$ and $v^{(q)}\left(x_0+(\underline{1}-i)h\right)$ up to terms involving $h^{s-\bar{s}}$ and $h^{s-|q|}$, respectively. Then (5) takes the form

$$\Delta = \sum_{|\beta|=0}^{s-s_0} \frac{h^\beta t^\beta}{\beta!} v^{(\alpha+\beta)}\left(x_0\right) -$$

$$- \sum_{|q|=0}^{s} \sum_{i\in I} \sum_{|\beta|=0}^{s-|q|} \frac{h^{|\beta+q|-s_0}(\underline{1}-i)^\beta}{\beta!} v^{(\beta+q)}\left(x_0\right)\tilde{\omega}_q^{(\alpha)}(t+i) +$$

$$+ \left(s-s_0+1\right)h^{s-s_0+1} \int_0^1 \sum_{|\beta|=s-s_0+1} \frac{t^\beta}{\beta!} v^{(\alpha+\beta)}\left(x_0+hzt\right)(1-z)^{s-s_0}dz -$$

$$- \sum_{|q|=0}^{s} (s-|q|+1)h^{s-|q|+1} \sum_{i\in I} \left[\int_0^1 \sum_{|\beta|=s-|q|+1} \frac{(\underline{1}-i)^\beta}{\beta!} \times\right.$$

$$\left. \times\, v^{(\beta+q)}\left(x_0+(1-i)hz\right)(1-z)^{s-|q|}dz\right]\tilde{\omega}_q^{(\alpha)}(t+i) . \tag{6}$$

The first two sums vanish. This can be proved in the same way as in §1: putting $\alpha + \beta = \gamma$ in the first sum and $\beta + q = \gamma$ in the second, we eliminate β. Then these two sums take the form

$$\sum_{|\gamma|=s_0}^{s} h^{|\gamma|-s_0} v^{(\gamma)}(x_0)\frac{t^{\gamma-\alpha}}{(\gamma-\alpha)!} - \sum_{|q|=0}^{s}\sum_{\substack{|\gamma|=|q|\\ \gamma\geq q}}^{s}\sum_{i\in I} h^{|\gamma|-s_0} v^{(\gamma)}(x_0)\frac{(\underline{1}-i)^{\gamma-q}}{(\gamma-q)!}\tilde{\omega}_q^{(\alpha)}(t+i)\ ,$$

or, changing the summation order in the second, the form

$$\sum_{|\gamma|=s_0}^{s} h^{|\gamma|-s_0} v^{(\gamma)}(x_0)\frac{t^{\gamma-\alpha}}{(\gamma-\alpha)!} - \sum_{|\gamma|=0}^{s} h^{|\gamma|-s_0} v^{(\gamma)}(x_0)\sum_{q\leq\gamma}\sum_{i\in I}\frac{(\underline{1}-i)^{\gamma-q}}{(\gamma-q)!}\tilde{\omega}_q^{(\alpha)}(t+i)\ .$$

The last expression is equal to zero by the fundamental completeness conditions (9.8) of Chapter II.

It now remains to examine the last two sums in (6), which correspond to the remainders of the Taylor series expansions. It is easily seen that they are cominated by

$$Ch^{s-s_0+1}\|v\|_{C^{(s+1)}(\bar{Q})}\ ,\quad C = \text{const.},$$

and this yields

$$\|v-v^h\|_{C^{(\bar{s})}(\bar{Q})} \leq C_1 h^{s-\bar{s}+1}\|v\|_{C^{(s+1)}(\bar{Q})}\ ,\quad C_1 = \text{const.} \tag{8}$$

In particular, the inequality (8) is valid for the function $\bar{u}_h = v_h$, introduced at the beginning of this section. So

$$\left\|v_h-v_h^h\right\|_{p,\bar{s}} \leq C_2\left\|v_h-v_h^h\right\|_{C^{(\bar{s})}(\bar{Q})} \leq C_3 h^{s-\bar{s}+1}\|v_h\|_{C^{(s+1)}(\bar{Q})}\ , \tag{9}$$

where

$$v_h^h(x) = \sum_{|q|=0}^{s}\sum_{j\in\hat{J}^h} h^q v_h^{(q)}\big((j+\underline{1})h\big)\tilde{\omega}_q\left(\frac{x}{h}-j\right)\ . \tag{10}$$

Duplicating the arguments of §3, they now yield the inequality (2).

§6. Strengthened Fundamental Completeness Conditions

THEOREM. *Assume that given primitive functions of dimension m and degree $s-1$ satisfy the following "strengthened fundamental completeness condition"*

$$\sum_{|q|=0}^{s-1}\sum_{i\in I}\frac{(\underline{1}-i)^{\gamma-q}}{(\gamma-q)!}\omega_q(t+i) = \frac{t^\gamma}{\gamma!}\ ,\quad |\gamma|\leq\tilde{s}\ , \tag{1}$$

where $\tilde{s} > s$ is some positive integer. For $u\in W_p^{(\tilde{s}+1)}(\Omega)$, let u be extended to $\mathbb{R}_m$ so that $u\in W_p^{(\tilde{s}+1)}(\mathbb{R}_m)$. Then there exist coefficients $a_{qj}^{(h)}$ such that

$$\left\| u - \sum_{|q|=0}^{s-1} \sum_{j\in J^h} a_{qj}^{(h)} \omega_q\left(\frac{x}{h} - j\right)\right\|_{p,\bar{s}} \le Ch^{\tilde{s}-\bar{s}+1}\|u\|_{p,\tilde{s}+1} , \quad 0 \le \bar{s} \le s . \tag{2}$$

The proof follows the pattern of §1 and §3. It suffices to consider functions of $C^{(\tilde{s}+1)}(\mathsf{R}_m)$ with compact support, and to prove the estimate

$$\|u-u^h\|_{C^{(\bar{s})}} \le C\|u\|_{C^{(\tilde{s}+1)}} h^{\tilde{s}-\bar{s}+1} \tag{3}$$

where $J^h(x)$ is defined by (1.2) of Chapter II. If $u \in W_p^{(\tilde{s}+1)}(\Omega)$, then we can apply the estimate (3) to the averaged function and repeat the argument of §3. The estimate (3) is proved as before: we put the support of u in a cube Q with sides parallel to the coordinate axes, choose h smaller than the distance between ∂Q and supp u , construct the grid with grid spacing h , take the smaller cube Q_{j_0} with lower vertex $x_0 = j_0 h$ and consider the difference

$$D^\alpha u(x) - D^\alpha u^h(x) = u^{(\alpha)}\left(x_0+th\right) - \sum_{|q|=0}^{s-1} \sum_{i\in I} h^{|q|-|\alpha|} u^{(q)}\left(x_0+(\underline{1}-i)h\right)\omega_q^{(\alpha)}(t+i) , \tag{4}$$

with $|\alpha| \le s$. Set $x = x_0 + th$, with $\underline{0} \le t \le \underline{1}$. We develop the Taylor series expansions for $u^{(\alpha)}\left(x_0+th\right)$ and $u^{(q)}\left(x_0+(\underline{1}-i)h\right)$,

$$u^{(\alpha)}\left(x_0+th\right) = \sum_{|\beta|=0}^{\tilde{s}-|\alpha|} \frac{h^\beta t^\beta}{\beta!} u^{(\alpha+\beta)}\left(x_0\right) + $$
$$+ (\tilde{s}-|\alpha|+1)h^{\tilde{s}-|\alpha|+1} \int_0^1 \sum_{|\beta|=\tilde{s}-|\alpha|+1} \frac{t^\beta}{\beta!} u^{(\alpha+\beta)}\left(x_0+hzt\right)(1-z)^{\tilde{s}-|\alpha|}dz ,$$

$$u^{(q)}\left(x_0+(\underline{1}-i)h\right) = \sum_{|\beta|=0}^{\tilde{s}-q} \frac{h^\beta(\underline{1}-i)^\beta}{\beta!} u^{(\beta+q)}\left(x_0\right) + $$
$$+ (\tilde{s}-|q|+1) \int_0^1 \sum_{|\beta|=\tilde{s}-|q|+1} \frac{h^{\tilde{s}-|q|+1}(\underline{1}-i)^\beta}{\beta!} u^{(\beta+q)}\left(x_0+hz(\underline{1}-i)\right)(1-z)^{\tilde{s}-|q|}dz .$$

We substitute this into (4). The remainders yield a term of order $O\left(h^{\tilde{s}-|\alpha|+1}\right)$, while the principal terms vanish. This can be proved as in §1. Hence, $|D^\alpha u(x)-D^\alpha u^h(x)| = O\left(h^{\tilde{s}-|\alpha|+1}\right)$. Taking the supremum of the left hand side and summing with respect to α , $|\alpha| \le \tilde{s}$, we derive (3).

Consequently, we can obtain a higher order approximation without an increase in the degree of the primitive system, if the system is required to satisfy additional fundamental completeness conditions for $|\gamma| > s$. Naturally, the question arises as to whether such system exists. We now examine existence for one-dimensional systems (see Mikhlin [18]).

If $m = 1$, then the strengthened fundamental completeness conditions take the form

$$\sum_{\substack{q=0\\ q\le\gamma}}^{s-1} \sum_{i=0}^{1} \frac{(1-i)^{\gamma-q}}{(\gamma-q)!}\,\omega_q(t+i) = \frac{t^\gamma}{\gamma!}\ , \quad 0 \le t \le 1\ , \quad 0 \le \gamma \le \tilde{s}\ . \tag{5}$$

The relations (5) can be regarded as defining $(\tilde{s}+1)$ equations in the $2s$ unknowns $\omega_q(t)$, $\omega_q(t+1)$, $0 \le q \le s-1$, $0 \le t \le 1$. We put $\tilde{s} = 2s - 1$, then the number of equations equals to the number of unknowns. We rewrite (5) in the form

$$\omega_\gamma(t+1) + \sum_{q=0}^{\gamma} \frac{\omega_q(t)}{(\gamma-q)!} = \frac{t^\gamma}{\gamma!}\ , \quad 0 \le \gamma \le s-1\ , \tag{6}$$

$$\sum_{q=0}^{s-1} \frac{\omega_q(t)}{(\gamma-q)!} = \frac{t^\gamma}{\gamma!}\ , \quad s \le \gamma \le 2s-1\ . \tag{7}$$

Using (6) we find all functions $\omega_q(t+1)$, once the $\omega_q(t)$ have been determined from the linear algebraic system (7) of order s . We prove that the determinant of the system (7) is non-zero. We denote it by D :

$$D = \begin{vmatrix} \frac{1}{s!} & \frac{1}{(s-1)!} & \frac{1}{(s-2)!} & \cdots & \frac{1}{1!} \\ \frac{1}{(s+1)!} & \frac{1}{s!} & \frac{1}{(s-1)!} & \cdots & \frac{1}{2!} \\ \frac{1}{(s+2)!} & \frac{1}{(s+1)!} & \frac{}{s!} & \cdots & \frac{1}{3!} \\ \cdots & \cdots & \cdots & \cdots & \cdots \\ \frac{1}{(2s-1)!} & \frac{1}{(2s-2)!} & \frac{1}{(2s-3)!} & \cdots & \frac{1}{s!} \end{vmatrix} = \frac{\prod_{k=1}^{s-1} k!}{\prod_{k=s}^{2s-1} k!} \begin{vmatrix} 1 & C_s^1 & C_s^2 & \cdots & C_s^{s-1} \\ 1 & C_{s+1}^1 & C_{s+1}^2 & \cdots & C_{s+1}^{s-1} \\ 1 & C_{s+2}^1 & C_{s+2}^2 & \cdots & C_{s+2}^{s-1} \\ \cdots & \cdots & \cdots & \cdots & \cdots \\ 1 & C_{2s-1}^1 & C_{2s-1}^2 & \cdots & C_{2s-1}^{s-1} \end{vmatrix} .$$

We now evaluate the determinant on the right hand side of this last expression, denoting it by D_0 . We subtract from each row of the determinant D_0 (starting from the bottom) the preceding row and use the identity $\binom{k}{n} = \binom{k}{n-1} + \binom{k-1}{n-1}$. As a result we obtain

$$D_0 = \begin{vmatrix} 1 & C_s^1 & C_s^2 & \cdots & C_s^{s-1} \\ 0 & 1 & C_s^1 & \cdots & C_s^{s-2} \\ 0 & 1 & C_{s+1}^1 & \cdots & C_{s+1}^{s-2} \\ \cdots & \cdots & \cdots & \cdots & \cdots \\ 0 & 1 & C_{2s-2}^1 & \cdots & C_{2s-2}^{s-2} \end{vmatrix} = \begin{vmatrix} 1 & C_s^1 & \cdots & C_s^{s-2} \\ 1 & C_{s+1}^1 & \cdots & C_{s+1}^{s-2} \\ \cdots & \cdots & \cdots & \cdots \\ 1 & C_{2s-2}^1 & \cdots & C_{2s-2}^{s-2} \end{vmatrix} .$$

Repeating the same argument for the resulting determinant we obtain finally that

$$D_0 = \begin{vmatrix} 1 & C_s^1 \\ \\ 1 & C_{s+1}^1 \end{vmatrix} = 1$$

and, hence,

$$D = \prod_{k=1}^{s-1} k! \ / \prod_{k=s}^{2s-1} k! \neq 0 .$$

Consequently, the equations (7) uniquely define the functions $\omega_q(t)$, $0 \le t \le 1$, $0 \le q \le s-1$. Obviously, these functions are polynomials of degree not greater than $2s - 1$. We prove that the following identities hold for them:

$$\omega_q^{(\alpha)}(0) = 0 , \quad \omega_q^{(\alpha)}(1) = \delta_{\alpha q} , \quad 0 \le \alpha, q \le s-1 . \tag{8}$$

We differentiate (7) α times, $0 \le \alpha \le s-1$:

$$\sum_{q=0}^{s-1} \frac{\omega_q^{(\alpha)}(t)}{(\gamma-q)!} = \frac{t^{\gamma-\alpha}}{(\gamma-\alpha)!} , \quad s \le \gamma \le 2s-1 . \tag{9}$$

The determinant of the system (9) is equal to $D \neq 0$. Setting $t = 0$ in (9), we obtain the new system

$$\sum_{q=0}^{s-1} \frac{\omega_q^{(\alpha)}(0)}{(\gamma-q)!} = 0 , \quad s \le \gamma \le 2s-1 ,$$

with the same determinant and with the obvious solution $\omega_q^{(\alpha)}(0) = 0$. Since $D \neq 0$, this solution is unique which establishes the first part of (8). Setting $t = 1$ in (9), we obtain the new system

$$\sum_{q=0}^{s-1} \frac{\omega_q^{(\alpha)}(1)}{(\gamma-q)!} = \frac{1}{(\gamma-\alpha)!} , \quad s \le \gamma \le 2s-1 ,$$

with the obvious and unique solution $\omega_q^{(\alpha)}(1) = \delta_{\alpha q}$. Thus, the second part of (8) also holds. Hence, it follows (see §6 of Chapter II) that the functions $\omega_q(t+1)$, $0 \le t \le 1$, also satisfy the necessary conditions

$$\omega_q^{(\alpha)}(1) = \delta_{\alpha q} , \quad \omega_q^{(\alpha)}(2) = 0 , \quad 0 \le \alpha, q \le s-1 . \tag{10}$$

Thus, if $\tilde{s} = 2s - 1$, there exist a primitive system of degree $s - 1$, satisfying the fundamental completeness conditions (5). If $0 \le t \le 1$, then they are polynomials of degree not greater than $2s - 1$ satisfying (8). But such polynomials are uniquely defined: they have been constructed in §6 of Chapter II. Hence, it follows that the primitive functions defined by (6.12), (6.13) and (6.17) of Chapter II (in particular, the primitive functions of Table 1, §6, Chapter II) yield the highest possible order of approximation: *if* $u \in W_p^{(\tilde{s}+1)}(0, 1)$, $s \le \tilde{s} \le 2s-1$, $0 \le \bar{s} \le s$,

$$\|u-u^h\|_{p,\bar{s}} \le C\|u\|_{p,\tilde{s}+1} h^{\tilde{s}-\bar{s}+1} , \quad C = \text{const.}, \tag{11}$$

where

$$u^h(x) = \sum_{q=0}^{s-1} \sum_{j=-1}^{2n-1} h^q u^{(q)}((j+1)h)\omega_q\left(\frac{x}{h} - j\right) .$$

No primitive system exists which satisfies the strengthened fundamental completeness conditions for $\tilde{s} > 2s-1$. Indeed, these relations, when rewritten for values of γ such that $0 \le \gamma \le 2s-1$, uniquely define the functions $\omega_q(t)$, $\omega_q(t+1)$, $0 \le t \le 1$, to be polynomials of degree not greater than $2s - 1$. No linear combination of these functions can equal $t^\gamma/\gamma!$, if $\gamma > 2s-1$.

We do not analyse in detail the case when $m > 1$. We only note that products of the one-dimensional primitive functions studied above yield m-dimensional primitive systems, for which the same order of approximation as that given in (11) holds.

§7. Some General Considerations

In this section, we derive estimates of the form

$$\|u-v^h\|_{p,\bar{s}} \le C\|u\|_{p,\tilde{s}+1} h^{\tilde{s}-\bar{s}+1} , \quad \tilde{s} \ge s \ge \bar{s} . \tag{1}$$

We consider (1) for functions $u \in W_p^{(\tilde{s}+1)}(Q)$, where Q is the cube $\underline{0} \le x \le \underline{a}$, and $u \equiv 0$ near ∂Q . Here

$$v^h(x) = \sum_{|q|=0}^{s-1} \sum_{j\in\mathfrak{J}^h} h^q v^{(q)}((j+\underline{1})h)\omega_q\left(\frac{x}{h} - j\right) , \tag{2}$$

where $v(x)$ denotes the averaging of $u(x)$ with averaging radius h . If the primitive functions are products, then we must sum over $|q|$ from 0 to s in (2). This does not influence the following reasonings.

For fixed h , the values of $v^{(q)}((j+1)h)$ are bounded functionals of u with respect to the norm of $W_p^{(\bar{s})}(Q)$. Consider a set of functions, which is bounded with respect to the norm of $W_p^{(\tilde{s}+1)}(Q)$. By the imbedding theorem, this set is bounded with respect to the norm of $W_p^{(\bar{s})}(Q)$. If h is fixed, then the corresponding set of functions $v^{(h)}(x)$ is compact in $W_p^{(\bar{s})}(Q)$, $\bar{s} \le \tilde{s}$. In fact, it is the compact ε-net for the set under consideration with $\varepsilon = CMh^{\tilde{s}-\bar{s}+1}$, $M = \sup\|u\|_{p,\tilde{s}+1}$.

The concept of imbedding can be formulated in a more general way. Let B_1 and B_2 be Banach spaces and B_1 be imbedded in B_2 such that the imbedding operator is bounded; that is, if $u \in B_1$, then $u \in B_2$ and there exists a constant μ such that

$$\|u\|_2 \le \mu\|u\|_1 , \quad \forall u \in B_1 , \quad \|\cdot\|_k = \|\cdot\|_{B_k} , \quad k = 1, 2 . \tag{3}$$

For each element $u \in B_1$, we identify one and only one element $v^h \in B_2$, with a positive value of h. We assume that, for fixed h, the operator T_h, transforming u into v^h, is finite-dimensional;

$$T_h u = v^h = \sum_{n=1}^{N} l_{nh}(u)\varphi_{uh} , \quad \varphi_{uh} \in B_2 . \tag{4}$$

Since the functionals l_{nh} are bounded in B_1, the operator T_h is a completely continuous mapping from B_1 into B_2.

We now suppose that the following estimate holds:

$$\|u-v^h\|_2 \le \gamma(h)\|u\|_1 \tag{5}$$

where $\gamma(h)$ is a monotone function, which tends to zero along with h. Let M be a bounded set in B_1 : $\forall u \in M$, $\|u\|_1 \le a \le$ const. Then the set $T_h M$ is compact in B_2. Putting $a\gamma(h) = \varepsilon$, we see that the set M has a compact ε-net in B_2 for any sufficiently small ε and, hence, M is compact in B_2. This means that B_1 is imbedded in B_2 and the imbedding operator is completely continuous. So the following theorem holds.

THEOREM. *If* B_1 *is boundedly imbedded in* B_2 *and the estimate* (5) *holds with* $\gamma(h) \downarrow 0$, *as* $h \downarrow 0$, *then* B_1 *is completely continuously imbedded in* B_2.

In particular, it follows from this theorem that the estimate

$$\|u-v^h\|_{p,s} \le \|u\|_{p,s}\gamma(h) , \quad \gamma(h) \downarrow 0 , \quad h \downarrow 0 ,$$

can not be valid.

§8. A More General Class of Primitive Systems

Above, we examined two classes of primitive systems: "the normal", described in §4, Chapter I, and the product (§9, Chapter I). These two classes have some general properties, which we now list:

(1) A primitive system contains a finite number of functions.

(2) The supports of the primitive functions are compact; more specifically, these supports are contained in the cube $\underline{0} \le t \le \underline{2}$.

(3) The primitive functions belong to the intersection of $C^{(s-1)}(\mathbb{R}_m)$ and $W_p^{(s)}(\mathbb{R}_m)$, where s and p are given, with s a positive integer and p such that $1 \le p \le \infty$.

(4) Let $\omega_l(t)$ be a primitive function, then there exists a definite multi-index α such that $\omega_l^{(\alpha)}(1) = 1$ and $\omega_l^{(\beta)}(1) = 0$ if $\beta \ne \alpha$ and the value of $|\beta|$ is not very great.

(5) If we enumerate in some order the primitive functions to form a primitive system consisting of the functions $\omega_l(t)$, $l = 1, 2, \ldots, r$, $r \ge 1$, then the corresponding coordinate functions are constructed to be

$$\varphi_{lj}^{(h)}(x) = \omega_l\left(\frac{x}{h} - j\right) , \tag{1}$$

where j is any integer vector.

(6) There exist "fundamental completeness conditions", which are sufficient (in some cases their necessity can be established) to guarantee that the system (1) is complete in $W_p^{(s)}(\Omega)$, where Ω denotes a bounded domain in $\mathbb{R}_m$. In some situations, it is necessary to assume that $\partial\Omega \in C^{(0,1)}$.

Below, the above properties (1)-(6) of the examined primitive system will be of central importence. The specific characterization of the primitive system will not be so important. In particular, the theorems and results of Chapters VII-IX are usually formulated for "normal" primitive systems, but they are also valid for any systems with properties (1)-(6), including product systems.

CHAPTER IV

PRIMITIVE FUNCTIONS WITH WIDE SUPPORT

§1. Definitions

In the present chapter we examine more complicated primitive systems. As will be shown below, the use of such systems can yield higher order approximations than systems studied previously.

We take natural numbers s and k and multi-indices q and r satisfying the inequalities

$$|q| \leq s-1 \ , \quad \underline{0} \leq r \leq \underline{k}-\underline{1} \ . \tag{1}$$

We call functions $\omega_{qr}(t)$ *primitive*, if they satisfy the following conditions (compare §4 of Chapter I)

$$\omega_{qr} \in C^{(s-1)}(\mathbb{R}_m) \cap W_p^{(s)}(\mathbb{R}_m) \ , \tag{2}$$

$$\operatorname{supp} \omega_{qr} \subset \{t : \underline{0} \leq t \leq 2\underline{k}\} \ , \tag{3}$$

$$\omega_{qr}^{(\alpha)}(\underline{1}+l+ki) = \begin{cases} 0 & , \quad i \neq \underline{0} \ , \quad i \in I \ , \\ \delta_{\alpha q}\delta_{lr} \ , & i = \underline{0} \ , \end{cases} \tag{4}$$

$|\alpha|$, $|q| \leq s-1$, $\underline{0} \leq l$, $r \leq \underline{k}-1$.

It follows from (2) and (3) that on the boundary B of the cube $\underline{0} \leq t \leq 2\underline{k}$, the primitive functions satisfy the first boundary value problem conditions

$$\omega_{qr}^{(\gamma)}(t) = 0 \ , \quad \forall t \subset B \ , \quad |\gamma| \leq s-1 \ . \tag{5}$$

Coordinate functions are constructed from the primitives using the following formula (similar to (4.4) of Chapter I):

$$\varphi^{(h)}_{qrj}(x) = \omega_{qr}\left(\frac{x}{h} - kj\right) , \tag{6}$$

where j is an arbitrary integer vector.

Let J be a finite set of vectors j and

$$v(x) = \sum_{|q|=0}^{s-1} \sum_{r=\underline{0}}^{\underline{k-1}} \sum_{j\in J} a_{qrj}\varphi^{(h)}_{qrj}(x) = \sum_{|q|=0}^{s-1} \sum_{r=\underline{0}}^{\underline{k-1}} \sum_{j\in J} a_{qrj}\omega_{qr}\left(\frac{x}{h} - kj\right) . \tag{7}$$

We apply to both sides of (7) the operator D^{α} , $|\alpha| \le s-1$, and set $x = \left(kj_0+r_0+\underline{1}\right)h$, where $j_0 \in J$ and $\underline{0} \le r_0 \le \underline{k-1}$. Using (3)-(5), we find

$$a_{\alpha r_0 j_0} = h^{\alpha} v^{(\alpha)}\left(\left(kj_0+r_0+\underline{1}\right)h\right) \tag{8}$$

and, hence

$$v(x) = \sum_{|q|=0}^{s-1} \sum_{r=\underline{0}}^{\underline{k-1}} \sum_{j\in J} h^{q} v^{(q)}\left((kj+r+\underline{1})h\right)\omega_{qr}\left(\frac{x}{h} - kj\right) . \tag{9}$$

It follows that the primitive functions of the present chapter are the same as the primitive functions of Chapter I, used to solve the Hermite interpolation problem. Let $u \in C^{(s)}(\bar{\Omega})$ where, as everywhere at this monograph, Ω denotes a bounded domain in R_m . We construct a grid of "small" cubes with sides of length h and a grid of "larger" cubes with sides of length $2kh$. We consider the grid domain $\hat{\Omega}^h \subset \Omega$, which consists of the union of larger cubes lying inside Ω , and denote by $\hat{J}^h$ the set of labels of the lower vertices of the larger cubes which define $\hat{\Omega}^h$. We require the function $u^h(x)$ which, along with its derivatives up to and including order α , $|\alpha| \le s-1$, coincides with $u(x)$ at the lower vertices of the small cubes lying inside $\hat{\Omega}^h$. It is clear from (9) that the corresponding interpolatory function can be defined as

$$u^h(x) = \sum_{|q|=0}^{s-1} \sum_{r=\underline{0}}^{\underline{k-1}} \sum_{j\in \hat{J}^h} h^{q} u^{(q)}\left((kj+r+\underline{1})h\right)\omega_{qr}\left(\frac{x}{h} - kj\right) . \tag{10}$$

The primitive functions of §4, Chapter I, are a particular case of the functions to be constructed in this section and in fact correspond to the special case when $k = 1$. As in Chapter I, the set of primitive functions will be called the primitive system; and the number $2k$ will be called the *width* of the primitive system's support. When $k = 1$, we shall refer to "primitive system with narrow support", and when $k > 1$ to "primitive system with wide support".

We now study in detail the one-dimensional systems with wide support. The corresponding multi-dimensional systems can be studied in an analogous way.

§2. Fundamental Completeness Conditions for One-Dimensional Systems

It is assumed that the primitive functions satisfy the conditions listed in the preceding section. In the one-dimensional case, they are

$$\omega_{qr} \in W_p^{(s)}(R_1) \ , \quad 0 \le q \le s-1 \ , \quad 0 \le r \le k-1 \ , \tag{1}$$

$$\operatorname{supp} \omega_{qr} \subset [0,\ 2k] \ , \tag{2}$$

$$\omega_{qr}^{(\alpha)}(1+l+ki) = \begin{cases} 0 & , \quad i = 1 \ , \\ \delta_{\alpha q}\delta_{lr} & , \quad i = 0 \ , \end{cases} \tag{3}$$

$0 \le \alpha,\ q \le s-1$, $0 \le l,\ r \le k-1$.

The boundary conditions at points $t = 0$ and $t = 2k$ are

$$\omega_{qr}^{(\alpha)}(0) = \omega_{qr}^{(\alpha)}(2k) = 0 \ , \quad 0 \le \alpha \le s-1 \ . \tag{4}$$

In the formula (1.6), which defines the coordinate functions, j now denotes an arbitrary integer.

It is not difficult to derive fundamental completeness conditions which are sufficient to guarantee that the coordinate system, constructed from the primitive system $\omega_{qr}(t)$, is complete in $W_p^{(s)}(0,\ 1)$. Suppose $u \in C^{(s)}(0,\ 1)$ and $\operatorname{supp} u \subset (0,\ 1)$. We take an integer n , put $h = 1/2nk$ and set

$$u^h(x) = \sum_{q=0}^{s-1}\sum_{r=0}^{k-1}\sum_{j=-1}^{2n-1} h^q u^{(q)}\big((kj+r+1)h\big)\omega_{qr}\left(\frac{x}{h} - kj\right) \ . \tag{5}$$

Consider

$$\int_0^1 \left|\frac{d^\alpha u(x)}{dx^\alpha} - \frac{d^\alpha u^h(x)}{dx^\alpha}\right|^p dx = \sum_{j_0=0}^{2n-1}\sum_{r_0=0}^{k-1}\int_{(kj_0+r_0)h}^{(kj_0+r_0+1)h}\left|\frac{d^\alpha u(x)}{dx^\alpha} - \frac{d^\alpha u^h(x)}{jx^\alpha}\right|^p dx$$

$$\overset{\text{def}}{=\!=} \sum_{j_0=0}^{2n-1}\sum_{r_0=0}^{k-1} F_{j_0 r_0} \ , \quad 0 \le \alpha \le s \ . \tag{6}$$

We examine the terms of (6) which do not vanish identically on the interval $\big((kj_0+r_0)h,\ (kj_0+r_0+1)h\big)$; that is, are non-zero under the integral sign in $F_{j_0 r_0}$.

These are the terms for which $0 \le \frac{x}{h}\ \ kj \le 2k$. We set $\frac{x}{h} = kj_0 + r_0 + t$, $0 < t < 1$. Then $\frac{x}{h} - kj = (j_0-j)k + r_0 + t$ and as a result $0 \le k(j_0-j)+r_0+t \le 2k$. Obviously, the last inequality holds if and only if $j_0 - j = i$, where i is either 0 or 1 . Now, (5) can be written in the form

$$u^h(x) = \sum_{q=0}^{s-1} \sum_{r=0}^{k-1} \sum_{i=0}^{1} h^q u^{(q)}\big(x_0+(r-r_0-ki+1)h\big)\omega_{qr}\big(t+r_0+ki\big) \ , \quad x_0 = \big(kj_0+r_0\big)h \ .$$

Hence,

$$\frac{d^\alpha u(x)}{dx^\alpha} - \frac{d^\alpha u^h(x)}{dx^\alpha} = u^{(\alpha)}\big(x_0+th\big) - \sum_{q=0}^{s-1} \sum_{r=0}^{k-1} \sum_{i=0}^{1} h^{q-\alpha} u^{(q)}\big(x_0+(r-r_0-ki+1)h\big)\omega_{qr}^{(\alpha)}\big(t+r_0+ki\big) \ .$$

We replace $u^{(\alpha)}\big(x_0+th\big)$ by $u^{(\alpha)}\big(x_0\big)$, take the Taylor series expansion of $u^{(q)}\big(x_0+(r-r_0-ki+1)h\big)$ about the point x_0 up to terms of $O\big(h^{\alpha-q-1}\big)$ and eliminate the remainders. The eliminated terms on the r.h.s. are of $O(1)$; the remaining terms on the r.h.s. form a polynomial with argument h^{-1} . Using the same arguments as in Chapter II, we obtain that the expression (6) tends to zero with h if and only if this polynomial is identically equal to zero. In this way we have obtained the following completeness conditions for the mentioned primitive functions

$$\sum_{\substack{q=0\\ q\le\gamma}}^{s-1} \sum_{r=0}^{k-1} \sum_{i=0}^{1} \frac{\big(1+r-r_0-ki\big)^{\gamma-q}}{(\gamma-q)!}\,\omega_{qr}^{(\alpha)}\big(t+r_0+ki\big) = \frac{t^{\gamma-\alpha}}{(\gamma-\alpha)!} \ ,$$

$$0 \le t \le 1 \ , \quad 0 \le \gamma \le s \ , \quad 0 \le r_0 \le k-1 \ , \quad 0 \le \alpha \le s \ . \qquad (8)$$

In particular, when $\alpha = 0$, we have

$$\sum_{\substack{q=0\\ q\le\gamma}}^{s-1} \sum_{r=0}^{k-1} \sum_{i=0}^{1} \frac{\big(1+r-r_0-ki\big)^{\gamma-q}}{(\gamma-q)!}\,\omega_{qr}\big(t+r_0+ki\big) = \frac{t^{\gamma}}{\gamma!} \ ,$$

$$0 \le t \le 1 \ , \quad 0 \le \gamma \le s \ , \quad 0 \le r_0 \le k-1 \ . \qquad (9)$$

Since (8) can be derived from (9) by differentiation, the conditions (9) (which we shall refer to as *fundamental completeness conditions*) are necessary and sufficient for $\|u-u^h\|_{p,s} \xrightarrow[h\to 0]{} 0$, if $u \in C^{(s)}[0,\, 1]$ and $\operatorname{supp} u \subset (0,\, 1)$, where $\|\cdot\|_{p,s}$ denotes the norm in $W_p^{(s)}(0,\, 1)$.

Now let $u(x)$ denote an arbitrary function in $W_p^{(s)}(0,\, 1)$. We extend it to $\mathbb{R}_1$ so that $u \in W_p^{(s)}\big(\mathbb{R}_1\big)$ and take its Sobolev average. Applying the arguments of §3, Chapter II, we find that the fundamental completeness conditions (9) are sufficient for the coordinate system to be complete in $W_p^{(s)}(0,\, 1)$.

It is not difficult to derive strengthened fundamental completeness conditions. They have the same form as (9), but are satisfied for the values $\gamma \le \tilde{s}$, with

$\tilde{s} > s$. If the strengthened fundamental completeness conditions are satisfied, then, for any function $u \in W_p^{(\tilde{s}+1)}(0, 1)$, there exist coefficients $a_{qrj}^{(h)}$ such that

$$\|u-u^h\|_{p,\bar{s}} \le C\|u\|_{p,\tilde{s}+1} h^{\tilde{s}-\bar{s}+1} , \quad 0 \le \bar{s} \le \tilde{s} , \quad C = \text{const.} \tag{10}$$

where

$$u^h(x) = \sum_{q=0}^{s-1} \sum_{r=0}^{k-1} \sum_{j=-1}^{2kh-1} a_{qrj}^{(h)} \omega_{qr}\left(\frac{x}{h} - kj\right) . \tag{11}$$

This can be proved in a similar way to the results of §6, Chapter III.

§3. Example: The Parabolic Approximation

Consider the following special case: $k = 2$, $s = 1$, $s = 2$. Then q can only have the value $q = 0$, and r the two values $r = 0$ and $r = 1$. We aim to construct two functions $\omega_0(x) = \omega_{00}(x)$ and $\omega_1(x) = \omega_{01}(x)$ satisfying the fundamental completeness conditions which now have the form

$$\left.\begin{aligned} \omega_0(t) + \omega_0(2+t) + \omega_1(t) + \omega_1(2+t) &= 1 , \\ \omega_0(t) + \omega_0(2+t) + 2\omega_1(t) &= t , \\ \omega_0(t) + \omega_0(2+t) + 4\omega_1(t) &= t^2 , \end{aligned}\right\} \tag{1}$$

$$\left.\begin{aligned} \omega_0(1+t) + \omega_0(3+t) + \omega_1(1+t) + \omega_1(3+t) &= 1 , \\ -2\omega_0(3+t) + \omega_1(1+t) - \omega_1(3+t) &= t , \\ 4\omega_0(3+t) + \omega_1(1+t) + \omega_1(3+t) &= t^2 . \end{aligned}\right\} \tag{2}$$

In addition, the conditions (2.4) should be satisfied. They can be rewritten using matrix notation as

$$\begin{Vmatrix} \omega_0(0) & \omega_0(1) & \omega_0(2) & \omega_0(3) & \omega_0(4) \\ \omega_1(0) & \omega_1(1) & \omega_1(2) & \omega_1(3) & \omega_1(4) \end{Vmatrix} = \begin{Vmatrix} 0 & 1 & 0 & 0 & 0 \\ 0 & 0 & 1 & 0 & 0 \end{Vmatrix} . \tag{3}$$

The equations (1) define a system of three equations in the four unknowns $\omega_0(t)$, $\omega_0(2+t)$, $\omega_1(t)$, $\omega_1(2+t)$. In the same way, (2) defines a system of three equations in the four unknowns $\omega_0(1+t)$, $\omega_0(3+t)$, $\omega_1(1+t)$, $\omega_1(3+t)$. Obviously, choosing arbitrary functions for $\omega_0(t)$ and $\omega_0(1+t)$, which are continuous for $0 \le t \le 1$, we can determine the other unknowns from (1) and (2). The chosen functions must satisfy the corresponding conditions in (3); namely,

$$\omega_0(t)\Big|_{t=0} = 0 \ , \quad \omega_0(t)\Big|_{t=1} = \omega_0(1+t)\Big|_{t=0} = 1 \ , \quad \omega_0(1+t)\Big|_{t=1} = 0 \ .$$

It is then easily verified that the remaining conditions in (3) are also satisfied.

If we take $\omega_0(t) = t(2-t)$, $0 \le t \le 2$, or, equivalently,

$$\omega_0(t) = t(2-t) \ , \quad \omega_0(1+t) = 1 - t^2 \ , \quad 0 \le t \le 1 \ ,$$

then we find from (1) and (2) that

$$\omega_0(t) = \begin{cases} t(2-t) \ , & 0 \le t \le 2 \ , \\ 0 \ , & 2 \le t \le 4 \ , \end{cases}$$
$$\omega_1(t) = \begin{cases} \tfrac{1}{2}t(t-1) \ , & 0 \le t \le 2 \ , \\ \tfrac{1}{2}(t-3)(t-4) \ , & 2 \le t \le 4 \ . \end{cases} \tag{4}$$

If $t < 0$ or $t > 4$, then both these functions are equal to zero (see Figure 15). The primitive functions (4) yield a continuous approximating function composed of parabolic arcs. On envoking the approximation

$$\int_0^1 u(x)ds = \int_0^1 u^h(x)dx \ ,$$

we obtain Simpson's rule.

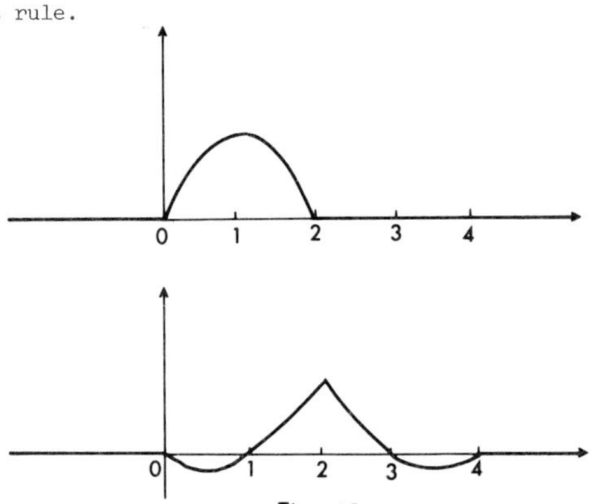

Fig. 15

NOTE. The estimate (2.10) does not depend on k . At the same time, as k increases, the primitive system becomes more complicated and hence the approximation $u^h(x)$ becomes more complicated. Consequently, k should be chosen as the smallest number for which the fundamental completeness conditions (2.9), considered as equations with unknowns $\omega_{qk}(t+r_0+ki)$, $0 \le t \le 1$, have a solution satisfying the conditions of §2. An analogous comment applies for the multidimensional case considered in the next section.

§4. Fundamental Completeness Conditions for Systems of Arbitrary Dimension

On this occasion, we shall only formulate the results, since the proofs follow the pattern of preceeding chapters and sections.

1. The fundamental completeness conditions take the form

$$\sum_{\substack{|q|=0 \\ q\leq\gamma}}^{s-1} \sum_{r=\underline{0}}^{\underline{k-1}} \sum_{i\in I} \frac{(1+r-r_0-ki)^{\gamma-q}}{(\gamma-q)!} \omega_{qr}(t+r_0+ki) = \frac{t^\gamma}{\gamma!} , \tag{1}$$

$$\underline{0} \leq t \leq \underline{1} , \quad |\gamma| \leq s , \quad \underline{0} \leq r_0 \leq \underline{k-1} . \tag{2}$$

2. For any function $u \in W_p^{(s)}(\Omega)$, where Ω is a finite domain in $\mathbb{R}_m$ and $\partial\Omega \in C^{(0,1)}$, there exists a finite set J of vectors j and coefficients $a_{qrj}^{(h)}$ such that $\|u-u^h\|_{p,s} \xrightarrow[h\to 0]{} 0$ where

$$u^h(x) = \sum_{|q|=0}^{s-1} \sum_{r=\underline{0}}^{\underline{k-1}} \sum_{j\in J} a_{qrj}^{(h)} \omega_{qr}\left(\frac{x}{h} - j\right) . \tag{3}$$

If $u \in C_0^{(s)}(\Omega)$, then we can take for J the set $\hat{J}^h$ of lower vertices of the larger cubes lying inside Ω and define the function $u^h(x)$ by (1.10).

3. The strengthened fundamental completeness conditions have the form (1), but they are satisfied for the larger set of values γ such that

$$|\gamma| \leq \tilde{s} , \quad \tilde{s} \geq s . \tag{4}$$

4. If $u \in W_p^{(\tilde{s}+1)}(\Omega)$ and $u(x)$ can be extended to $\mathbb{R}_m$ so that $u \in W_p^{(s+1)}(\mathbb{R}_m)$, then there exists a function $u^h(x)$ of the form (3) such that

$$\|u-u^h\|_{p,\bar{s}} \leq C\|u\|_{p,\tilde{s}+1} h^{\tilde{s}-\bar{s}+1} , \quad 0 \leq \bar{s} \leq s . \tag{5}$$

The construction of the primitive system of dimension $m > 1$ may appear to be difficult. In such situations, use can be made of tensor products of one-dimensional primitive functions (see §9, Chapter II) to construct product primitive systems with wide support. Construct one-dimensional primitive functions $\omega_{qr}(t)$ of degree $s - 1$ and width $2k$ which satisfy the conditions (2.1), (2.4), (2.9), with q, r and t being scalars. Now let q and r be multi-indices and t a point in $\mathbb{R}_m$ such that

$q = (q_1, \ldots, q_m)$, $r = (r_1, \ldots, r_m)$, $t = (t_1, \ldots, t_m)$. We set

$$\tilde{\omega}_{qr}(t) = \prod_{\mu=1}^{m} \omega_{q_\mu r_\mu}(t_\mu) \tag{6}$$

and consider only those functions (6) for which

$$\underline{0} \le r \le \underline{k-1} \ , \quad 0 \le |q| \le s \ , \quad s > 1 \ . \tag{7}$$

We note from (6), that the multi-index q only has the value $q = 0$ when $s = 1$. We construct the coordinate functions from the product primitive system using the formula

$$\varphi^{(h)}_{qrj}(x) = \tilde{\omega}_{qr}\left(\frac{x}{h} - kj\right) , \tag{8}$$

which is analogous to (1.6).

Repeating the arguments of §9, Chapter II, and §5, Chapter III, it is easy to verify that statements 2 and 4 of the present section continue to hold when the primitive functions $\omega_{qr}(x)$ are replaced by $\tilde{\omega}_{qr}(x)$.

CHAPTER V

APPROXIMATION IN ONE-DIMENSIONAL DEGENERATE NORMS

In Chapters V and VI, we shall examine approximation in degenerate norms; that is, weighted Sobolev norms which can tend to zero or infinity. We shall construct the approximation using primitive functions in the same manner as previously.

In the present chapter, we consider some special classes of functions. These are the solutions of first boundary value problem for degenerate ordinary differential equations of second order. The approximations will be examined using the energy norms defined by such problems. The basic content of the present chapter can be found in Mikhlin [12].

§1. The Formulation of the Problem

Consider the equation

$$-\frac{d}{dx}\left(\varphi(x)\frac{du}{dx}\right) + q(x)u = f(x) \ , \quad 0 < x < 1 \ , \quad f \in L_2(0, 1) \ , \tag{1}$$

in which $\varphi \in C[0, 1] \cap C^1(0, 1]$ and $q(x)$ is a measurable, bounded and non-negative function. Let $\varphi(0) = 0$, with $\varphi(x) > 0$ when $x > 0$, so that (1) is degenerate only at the point $x = 0$.

We distinguish between three distinct cases for (1) (see Mikhlin [2]):

(a) The integral

$$\int_0^1 \frac{dx}{\varphi(x)} \tag{2}$$

converges.

(b) The integral (2) diverges, but the integral

$$\int_0^1 \frac{x\,dx}{\varphi(x)} \tag{3}$$

converges.

(c) The integral (3) diverges.

The situation when

$$\varphi(x) = x^{\alpha}p(x) \ , \quad \alpha = \text{const} > 0 \ , \quad p \in C^{(1)}[0, 1] \ , \qquad p(x) \geq p_0 = \text{const.} > 0 \ , \tag{4}$$

is of particular interest. Here, (2) converges when $0 < \alpha < 1$ and diverges when $\alpha \geq 1$, while (3) converges when $0 < \alpha < 2$ and diverges when $\alpha \geq 2$.

The aim of the present chapter is to determine the order of approximation when the solution of (1) is approximated by a linear combination of coordinate functions $\varphi_{jh}(x) = \omega\left(\frac{x}{h} - j\right)$. We denote by ω the one-dimensional primitive function with narrow support, and degree zero (see §6, Chapter II):

$$\omega(t) = \begin{cases} t \ , & 0 \leq t \leq 1 \ , \\ 2-t \ , & 1 < t \leq 2 \ , \\ 0 \ , & \text{otherwise.} \end{cases} \tag{5}$$

The dependence of the order of approximation on $\varphi(x)$, $q(x)$ and $f(x)$ will be studied with respect to the energy metrics.

We denote by A_0 the operator which maps $L_2(0, 1)$ according to

$$\mathsf{A}_0 u = -\frac{d}{dx}\,\varphi(x)\,\frac{du}{dx} \ . \tag{6}$$

We take as the domain $\mathcal{D}(\mathsf{A}_0)$ of this operator the set of functions satisfying the conditions: $u(x)$ and $\varphi(x)u'(x)$ are absolutely continuous on any segment $[\delta, 1]$, $0 < \delta < 1$; if $\alpha < 1$, then these functions are continuous on the segment $[0, 1]$; $\mathsf{A}_0 u \in L_2(0, 1)$; if the integral (2) converges, then

$$u(0) = u(1) = 0 \ , \tag{7}$$

if the integral (2) diverges, then

$$u(1) = 0 \ . \tag{8}$$

We set $\mathsf{A}u = \mathsf{A}_0 u + qU$ and assume $f \in L_2(0, 1)$. Under these conditions, we formulate as the first boundary value problem for (1), the problem of solving $\mathsf{A}u = f$.

Below we shall use the following results (see Mikhlin [2], [3]).

If the integral (3) converges, then the operator A_0 (as well as the operator

A) is positive definite. It is also positive definite, if $\varphi(x)$ has the form (4) and $\alpha = 2$. If the integral (3) diverges, then the operator A_0 is positive definite when

$$q(x) \geq q_0 = \text{const} > 0 \ . \tag{T}$$

It is well known that, if A is positive definite, then the (operator) equation $Au = f$ has a unique generalised (weak) solution contained in the energy space H_A of A . This solution also belongs to $L_2(0, 1)$. In our case, H_A is the space of absolutely continuous functions on any segment $[\delta, 1]$, $0 < \delta < 1$, satisfying the condition (8) and giving a finite value to the integral (9) (see below). If the integral (2) converges, then these functions are continuous on the segment $[0, 1]$ and satisfy (7). The energy inner product and norm are defined by

$$[u, v]_A = \int_0^1 (\varphi u'v' + quv)dx \ ,$$
$$\|u\|_A^2 = \int_0^1 \left(\varphi u'^2 + qu^2\right) dx \ . \tag{9}$$

The problem $Au = f$ can be solved by the energy method using either the Ritz or variational-difference process. Hence, we need a system of coordinate functions which is complete with respect to the energy norm of (9). We shall show in §2 that this completeness property is valid for coordinate systems which are complete with respect to the energy norm of the non-degenerate operator B defined by

$$Bu = -\frac{d^2u}{dx^2} \ , \quad u(0) = u(1) = 0 \ . \tag{10}$$

We recall that

$$[u, v]_B = \int_0^1 u'v'dx \ , \quad \|u\|_B^2 = \int_0^1 u'^2 dx \ . \tag{11}$$

§2. On the Completeness of Coordinate Systems Which are Complete with Respect to Non-Degenerate Norms

(a) THE RITZ PROCESS

THEOREM 1. *Let the system* $\{\varphi_n(x)\}$ *contained in* $\overset{\circ}{W}{}_2^{(1)}(0, 1)$ *be complete with respect to the norm* (1.11). *Then it is complete with respect to the norm* (1.9).

Proof. Assume that the system $\{\varphi_n(x)\}$ is not complete in H_A . Then there exists a function $\sigma \in H_A$ such that $[\sigma, \varphi_n]_A = 0$, $n = 1, 2, \ldots$. Consider the set $\mathcal{D}$ of functions from $C^{(2)}[0, 1]$ with supports inside $(0, 1)$. Since the system $\{\varphi_n\}$ is complete in $\mathcal{D}$ with respect to (1.11), it is also complete in this set with respect to (1.9). Hence, it follows that, for any function $\eta \in \mathcal{D}$, $[\sigma, \eta]_A = 0$. Since $\mathcal{D} \subset \mathcal{D}(A)$, it follows that $(\sigma, A\eta) = [\sigma, \eta]_A = 0$; that is,

$$\int_0^1 \sigma(\xi)\left[-\frac{d}{d\xi}\varphi(\xi)\frac{d\eta}{d\xi} + q(\xi)\eta(\xi)\right]d\xi = 0 . \tag{1}$$

Now let a denote a sufficiently small scalar. As the function $\eta(\xi)$, we take the averaging kernel $\omega_\delta(|x-\xi|)$, $x \in [2a, 1]$, with averaging radius $\delta < a$. The identity (1) becomes $(\varphi\sigma')'_\delta = (q\sigma)_\delta$. Obviously, $\sigma \in L_2(2a, 1)$, so $(q\sigma)_\delta \xrightarrow[\delta\to 0]{} \sigma$ with respect to the norm in $L_2(2a, 1)$. Hence, the function $(\varphi\sigma')'_\delta$ has the same limit. On the other hand $\varphi\sigma' \in L_2(2a, 1)$ and $(\varphi\sigma')'_\delta \xrightarrow[\delta\to 0]{} \varphi\sigma'$ with respect to the norm in $L_2(2a, 1)$. Since the operator of generalised differentiation is closed, there exists a generalised derivative $(\varphi\sigma')' = q\sigma$. The generalised derivative for functions of one independent variable coincides almost everywhere with the usual derivative, so the function σ satisfies almost everywhere on $[2a, 1]$ the equation

$$-\frac{d}{dx}\varphi(x)\frac{d\sigma}{dx} + q(x)\sigma = 0 . \tag{2}$$

Since a is arbitrary small, the equation (2) is satisfied almost everywhere on the interval $(0, 1)$.

From (1.9) we have

$$\|\sigma\|_A^2 = \int_0^1 \left(\varphi\sigma'^2 + q\sigma^2\right)dx = \lim \int_\delta^1 \left(\varphi\sigma'^2 + q\sigma^2\right)dx .$$

We integrate this by parts. Using (2) and the boundary condition $\sigma(1) = 0$, we obtain

$$\|\sigma\|_A^2 = -\lim_{\delta\to 0}\varphi(\delta)\sigma(\delta)\sigma'(\delta) . \tag{3}$$

Suppose the integral (1.2) converges, then $\sigma(\delta) \xrightarrow[\delta\to 0]{} 0$. We integrate (2) over the range δ to 1:

$$\varphi(\delta)\sigma'(\delta) - \varphi(1)\sigma'(1) = \int_\delta^1 q(x)\sigma(x)dx .$$

Since the operator A is positive definite and $\sigma \in H_A$, $\sigma \in L_2(0, 1)$ and the last integral is finite for $\delta \to 0$. Hence the limit in (3) equals zero, which implies $\sigma(x) \equiv 0$.

Now assume that (1.2) diverges. From (2) we have

$$\frac{d\sigma}{dx} = -\frac{1}{\varphi(x)}\int_0^x q(t)\sigma(t)dt + \frac{C_0}{\varphi(x)} . \tag{4}$$

Since $\sigma \in L_2(0, 1)$, (4) has, for small values of x, order $O(\sqrt{x})$, so the principal order is given by the term $C_0/\varphi(x)$. But, if $C_0 \neq 0$, then $\|\sigma\|_A = \infty$. In fact,

$$\|\sigma\|_A^2 = \int_0^1 [\varphi(x)\sigma'^2(x)+q(x)\sigma^2(x)]dx .$$

For small values of x (let it be for $0 < x < \tau$), we have $|\sigma'(x)| \geq |C_0|/2\varphi(x)$, so

$$\|\sigma\|^2 \geq \frac{C_0^2}{4}\int_0^\tau \frac{dx}{\varphi(x)} = \infty .$$

Hence it follows that $C_0 = 0$ and thus

$$\frac{d\sigma}{dx} = -\frac{1}{\varphi(x)}\int_0^x q(t)\sigma(t)dt . \tag{5}$$

We now estimate (5):

$$\left|\int_0^x q(t)\sigma(t)dt\right| \leq \left\{\int_0^x q^2(t)dt\right\}^{\frac{1}{2}}\left\{\int_0^x \sigma^2(t)dt\right\}^{\frac{1}{2}} .$$

The function q is bounded, and σ^2 is summable; the first multiplier has order $O(\sqrt{x})$ and the second $O(1)$. Finally

$$|\sigma'(x)| = O\left(\frac{\sqrt{x}}{\varphi(x)}\right) . \tag{6}$$

Suppose the integral (1.3) converges. We derive an estimate for $|\sigma(x)|$. From (6) it follows that $|\sigma'(x)| \leq C\sqrt{x}/\varphi(x)$, $C = \text{const.}$ Hence

$$|\sigma(x)| \leq C\int_x^1 \frac{\sqrt{t}}{\varphi(t)}dt = C\int_x^1 \frac{1}{\sqrt{t}}\frac{tdt}{\varphi(t)} < \frac{C}{\sqrt{x}}\int_0^1 \frac{tdt}{\varphi(t)} = O\left(\frac{1}{\sqrt{x}}\right) .$$

By (3) we obtain that $\varphi(\delta)\sigma(\delta)\sigma'(\delta) = O(1)$. Thus the limit in (3) equals zero, and hence $\sigma(x) \equiv 0$.

It remains to consider the case when (3) diverges. From (3) it is clear that the value $\varphi(\delta)\sigma(\delta)\sigma'(\delta)$ has the non-positive limit $-\gamma = -\|\sigma\|_A^2$ when $\delta \to 0$. We prove that $\gamma = 0$. Let $\gamma > 0$. Since $\varphi(t)[\sigma^2(t)]' \xrightarrow[t\to 0]{} -2\gamma$, we have for a sufficiently small value of x_0 ,

$$\frac{d}{dt}\sigma^2(t) < -\frac{\gamma}{\varphi(t)}, \quad 0 < t \le x_0 .$$

Let $x < x_0$. Integrating over the range from x to x_0 , we find that

$$\sigma^2(x) > \sigma^2(x_0) + \gamma \int_x^{x_0} \frac{dt}{\varphi(t)} .$$

We integrate the last inequality over the range from zero to x_0 with respect to x :

$$\int_0^{x_0} \sigma^2(x)dx > x_0\sigma^2(x_0) + \gamma \int_0^{x_0} \left\{\int_x^{x_0} \frac{dt}{\varphi(t)}\right\}dx = x_0\sigma^2(x_0) + \gamma \int_0^{x_0} \frac{tdt}{\varphi(t)} .$$

If $\gamma > 0$, then the right hand side is infinite, which contradicts $\sigma \in L_2(0, 1)$. Finally, since $\gamma = \|\sigma\|_A^2 = 0$ and $\sigma(x) \equiv 0$, the required result follows.

(b) THE VARIATIONAL-DIFFERENCE PROCESS

For this process, we consider (see Chapter I) a sequence of finite-dimensional subspaces $\left\{H_A^{(n)}\right\}$ in H_A . The sequence $\left\{H_A^{(n)}\right\}$ is complete in H_A if, given $u \in H_A$ and $\varepsilon > 0$, one can find an N such that for any $n > N$ there exists an element $u^{(n)} \in H_A^{(n)}$ with

$$\|u-u^{(n)}\|_A < \varepsilon . \tag{7}$$

THEOREM 2. *Let* A *and* B *be the operators introduced in* §1 *and let all the subspaces* $\left\{H_A^{(n)}\right\}$ *belong to both energy spaces* H_A *and* H_B . *If this sequence is complete in* H_B , *then it is complete in* H_A .

Proof. We assume the contrary. Then there exist an element $\sigma_0 \in H_A$, a number $\varepsilon_0 > 0$ and a sequence $n_k \xrightarrow[k\to\infty]{} \infty$ such that $\left\|\sigma_0-u^{(n_k)}\right\|_A \ge \varepsilon_0$, $\forall u^{(n_k)} \in H_A^{(n_k)}$. We denote by η an arbitrary infinitely differentiable function on $[0, 1]$ with support contained in $(0, 1)$, and by η_n its projection on the subspace $H_A^{(n)}$ with respect to the norm in H_A . Then

$$\|\sigma_0-\eta\|_A \ge \|\sigma_0-\eta_{n_k}\|_A - \|\eta-\eta_{n_k}\|_A \ge \varepsilon_0 - \|\eta-\eta_{n_k}\|_A .$$

But $\|\eta-\eta_{n_k}\|_A$ tends to zero. Indeed, since the sequence $\left\{H_A^{(n)}\right\}$ is complete in

H_B , we have $\|\eta-\eta'_n\|_B \xrightarrow[n\to\infty]{} 0$, where η'_n denotes the projection of η on $H_A^{(n)}$ with respect to the norm in H_B . The metric H_A is weaker than H_B , consequently, $\|\eta-\eta'_n\|_A \xrightarrow[n\to\infty]{} 0$. Hence $\|\eta-\eta_{n_k}\|_A \le \|\eta-\eta'_n\|_A \xrightarrow[k\to\infty]{} 0$. Thus, we have $\|\sigma_0-\eta\|_A \ge \varepsilon_0$. We denote the closure of the set of functions η with respect to the norm in H_A by Y . Obviously, $\|\sigma_0-\zeta\|_A \ge \varepsilon_0$, $\forall\zeta \in Y$. If σ' denotes the projection of the element σ_0 on the subspace Y in H_A , then $[\sigma, \zeta]_A = 0$, $\sigma = \sigma_0 - \sigma'$, $\forall\zeta \in Y$. In particular, $[\sigma, \eta]_A = 0$ and the identity (1) holds. The rest follows in the same manner as in the proof of Theorem 1.

§3. Equations of Second Order with Weak Degeneracy

In §1, it was indicated that our aim is the study of variational-difference approximation with respect to energy norms generated by the degenerate problem (1.1), (1.7) or (1.1), (1.8). We shall compare the solution $u(x)$ with the peicewise linear approximation $u^h(x)$ which at the grid-point (except perhaps the grid-point $x = 0$) has the same ordinates as $u(x)$. Let $u_h(x)$ denote an approximation generated by some variational-difference method, and having the form of the piecewise linear approximation with the same grid points. Then $\|u-u_h\|_A \le \|u-u^h\|_A$. If we obtain an estimate for the right hand side of this inequality, then we have an estimate of the error for the approximate solution $u_h(x)$.

Below, we shall denote constants by C . We denote by $\|\cdot\|_r$, the norm in $L_r(0, 1)$; if $r = 2$, then we shall simply write $\|\cdot\|$.

In the present section, we consider the case when $\varphi(x)$ has the form (1.4) and $0 < \alpha < 1$. We note that the results of the present section can also be obtained by using the method of Gusman and Oghanes'an [1].

So, we consider the problem

$$-\frac{d}{dx}\, x^\alpha p(x)\, \frac{du}{dx} + q(x)u = f(x) , \quad 0 < x < 1 , \quad 0 < \alpha < 1 , \quad u(0) = u(1) = 0 \tag{1}$$

Let $u(x)$ be its solution. Since $f \in L_2(0, 1)$ and the operator defined by (1) is positive definite in $L_2(0, 1)$, $u(x)$ belongs to the same space and $\|u\| \le C\|f\|$. But then $g = (f-qu) \in L_2(0, 1)$ and $\|g\| \le C\|f\|$. Integrating, we find

$$\frac{du}{dx} = -\frac{1}{x^\alpha p(x)}\int_0^x g(t)dt + \frac{\gamma}{x^\alpha p(x)} , \quad \gamma = \text{const.} \tag{2}$$

By the Cauchy-Schwarz inequality, we obtain

$$\int_0^x g(t)dt = O(\sqrt{x})\|f\| \ .$$

Hence, it is obvious that the principal term in (2) is the second and $|u'(x)| \leq C\|f\|\cdot x^{-\alpha}$. Differentiation of (1) yields

$$u''(x) = -\left(x^{\alpha}p(x)\right)'\left(x^{\alpha}p(x)\right)^{-1}u'(x) - \left(x^{\alpha}p(x)\right)^{-1}g(x) \ . \tag{3}$$

The first term in (3) is dominated by $C\|f\|x^{-1-\alpha}$; the second term has the form $x^{-\alpha}g_1(x)$, where $g_1 = gp^{-1} \in L_2(0, 1)$.

We set $h = 1/2n$, $x_k = kh$ (n is an integer) and construct the piecewise linear approximation $u^h(x)$ with vertices $\left(x_k, u(x_k)\right)$, $k = 0, 1, \ldots, 2n$. We set

$$J = \int_0^1 x^{\alpha}p(x)|u'(x)-u^{h'}(x)|^2dx$$

$$= \sum_{k=0}^{2n-1} \int_{x_k}^{x_{k+1}} x^{\alpha}p(x)|u'(x)-u^{h'}(x)|^2dx \overset{\text{def}}{=\!=} \sum_{k=0}^{2n-1} J_k \ . \tag{4}$$

Putting $p = 2$ in (4.4) of Chapter III, we obtain

$$\int_{x_k}^{x_{k+1}} |u'(x)-u^{h'}(x)|^2dx \leq h^2 \int_{x_k}^{x_{k+1}} [u''(x)]^2dx$$

and, hence,

$$J_k \quad p_1(k+1)^{\alpha}h^{2+\alpha} \int_{x_k}^{x_{k+1}} [u''(x)]^2dx \ , \quad p_1 = \text{const.}$$

An estimate for $k \geq 1$ follows from (3),

$$\int_{x_k}^{x_{k+1}} [u''(x)]^2dx \leq 2C\|f\|^2 \int_{x_k}^{x_{k+1}} \frac{dx}{x^{2+2\alpha}} + 2\int_{x_k}^{x_{k+1}} \frac{g_1^2(x)}{x^{2\alpha}}\,dx \ .$$

Hence

$$\int_{x_k}^{x_{k+1}} [u''(x)]^2dx \leq \frac{2C\|f\|^2}{k^{2+2\alpha}h^{1+2\alpha}} + \frac{2}{k^{2\alpha}h^{2\alpha}} \int_{x_k}^{x_{k+1}} g_1^2(x)dx$$

and thus

$$J_k \le \frac{2Cp_1 h^{1-\alpha}}{k^{2+\alpha}} \left(\frac{k+1}{k}\right)^{\alpha} \|f\|^2 + \frac{2p_1 h^{2-\alpha}}{k^{\alpha}} \left(\frac{k+1}{k}\right)^{\alpha} \int_{x_k}^{x_{k+1}} g_1^2(x)dx$$

$$\le \frac{C_1 h^{1-\alpha}\|f\|^2}{k^2} + Ch^{2-\alpha} \int_{x_k}^{x_{k+1}} g_1^2(x)dx \ .$$

Taking the summation over k, we obtain $J \le J_0 + Ch^{1-\alpha}\|f\|^2$. We now estimate J_0:

$$J_0 = \int_0^h x^{\alpha} p(x) \left[u'(x) - \frac{u(h)}{h}\right]^2 dx \le 2p_1 \int_0^h x^{\alpha} u'^2(x)dx$$

$$+ \frac{2p_1}{1+\alpha} \frac{u^2(h)}{h^{1-\alpha}} \le Ch^{1-\alpha}\|f\|^2 + C\frac{u^2(h)}{h^{1-\alpha}} \ .$$

But

$$|u(h)| = \left|\int_0^h u'(t)dt\right| \le C\|f\| \int_0^h t^{-\alpha}dt = C\|f\|h^{1-\alpha} \ .$$

Hence $J_0 \le C\|f\|^2 h^{1-\alpha}$ and

$$\int_0^1 x^{\alpha} p(x) |u'(x) - u^{h\prime}(x)|^2 dx \le C\|f\|^2 h^{1-\alpha} \ . \tag{5}$$

It remains to estimate the integral

$$\int_0^1 q(x) |u(x) - u^h(x)|^2 dx \ .$$

Since the operator A_0 is assumed to be positive definite, it follows that

$$\int_0^1 q(x)|u(x)-u^h(x)|^2 dx \le q_1\|u-u^h\|^2 \le q_1 \varkappa \int_0^1 x^{\alpha}p(x)|u'(x)-u^{h\prime}(x)|^2 dx \le C\|f\|^2 h^{1-\alpha} \ , \tag{6}$$

$$\varkappa = \text{const} \ , \quad q_1 = \sup q(x) \ .$$

From (5) and (6) we finally obtain

$$\|u-u^h\|_A \le C\|f\|h^{(1-\alpha)/2} \ . \tag{7}$$

The estimate (7) has the precise order. This is a consequence of the following example (*cf.* Gusman and Oganesjan [1]). Let $u_0 \in C^{\infty}[0, 1]$, $u_0(1) = 0$ and $u_0(x) = x^{1-\alpha}/(1-\alpha)$, $0 \le x \le \frac{1}{2}$. Obviously,

$$f_0(x) = -\frac{d}{dx}\left(x^\alpha \frac{du_0}{dx}\right) \in C^{(\infty)}[0, 1]$$

and $f_0(x) = 0$, $0 \le x \le \frac{1}{2}$. So $u_0(x)$ can be considered as the solution of (1). We show that, for any piecewise linear approximation $u^h(x)$ passing through the points $(0, 0)$, $(1, 0)$ and having grid points kh , where the k are integers, the following inequality holds

$$\int_0^1 x^\alpha \left[u_0'(x) - u^{h\prime}(x)\right]^2 dx \ge c_0 h^{1-\alpha} \ , \quad c_0 = \text{const.} > 0 \ .$$

It is sufficient to verify that this inequality holds for the integral

$$\int_0^h x^\alpha \left(x^{-\alpha} - \beta\right)^2 dx \ ,$$

where β is an arbitrary constant. This integral has a minimum value when $\beta = (1+\alpha)h^{-\alpha}$. The value of the minimum is $\alpha^2 h^{1-\alpha}/(1-\alpha)$, from which the required result follows.

NOTE. The extimate (7) can be improved by changing the approximation. We set $x^{1-\alpha} = \tau$. Problem (1) is transformed into

$$-\frac{d}{d\tau} P(\tau) \frac{du}{d\tau} + Q(\tau)u = F(\tau) \ , \quad 0 < \tau < 1 \ , \quad u(0) = u(1) = 0 \ , \tag{8}$$

where

$$P(\tau) = (1-\alpha)p(x) \ , \quad Q(\tau) = \frac{x^\alpha q(x)}{1-\alpha} \ , \quad F(\tau) = \frac{x^\alpha f(x)}{1-\alpha} \ .$$

Problem (8) is non-degenerate. The coordinate functions $\omega\left(\frac{\tau}{h} - j\right)$, where ω is the function (1.5), give an order of approximation $O(h)\|f\|$ w.r.t. the energy norm. It remains to note that the energy norms (1) and (8) coincide. In fact, the square root of the last norm is

$$\int_0^1 \left[P(\tau)\left(\frac{du}{d\tau}\right)^2 + Q(\tau)u^2\right] d\tau \ . \tag{9}$$

Under the change of variables $\tau = x^{1-\alpha}$, the integral (9) transforms into the second integral in (1.9).

In the book by Varga [1], it is pointed out that, by applying the "singular" splines, one can obtain an order of approximation $O\left(h^{2-\alpha}\right)$ for the problem (1) w.r.t. the uniform norm. From the results of Chapter III, it follows that one can obtain an arbitrary high order of approximation, if the functions $P(\tau)$, $Q(\tau)$ and $F(\tau)$ are sufficiently smooth and if a primitive system of sufficiently high degree is used. We also note that the mentioned coordinate functions give an order $O(h^{2-\alpha})$, when $1 \le \alpha < 2$ (see §9 below).

§4. The Case $1 \le \alpha \le 2$

In this case, we must put $\gamma = 0$ in (3.2) otherwise the energy integral (1.9) diverges. Now it follows from (3.2) that $|u'(x)| \le C\|f\|\cdot x^{\frac{1}{2}-\alpha}$.

We derive an estimate for the second derivative. As before, (3.3) is valid. The first term on the r.h.s. is dominated by $C\|f\|x^{-(\alpha+\frac{1}{2})}$ and the second term, as in §3, has the form $x^{-\alpha}g_1(x)$, $g_1 \in L_2(0, 1)$, $\|g_1\| \le C\|f\|$. If $k \ge 1$, then

$$\int_{x_k}^{x_{k+1}} [u''(t)]^2 dt \le C\|f\|^2 \int_{x_k}^{x_{k+1}} \frac{dt}{t^{1+2\alpha}} + 2\int_{x_k}^{x_{k+1}} \frac{g_1^2(t)}{t^{2\alpha}}\, dt$$

$$\le \frac{C\|f\|^2}{k^{1+2\alpha}h^{2\alpha}} + \frac{2}{k^{2\alpha}h^{2\alpha}} \int_{x_k}^{x_{k+1}} g_1^2(t)dt .$$

Consider (3.4). We choose the piecewise linear approximation $u^h(x)$ as follows: it has vertices $(x_k, u(x_k))$ for $k = 1, 2, \ldots, 2n$, and $u^h(x) = u(x_1)$ for $x \in [0, h]$. If $k \ge 1$, then

$$J_k = \int_{x_k}^{x_{k+1}} x^\alpha p(x)|u'-u^{h'}|^2 dx \le p_1(k+1)^\alpha h^\alpha \int_{x_k}^{x_{k+1}} |u'-u^{h'}|^2 dx ,$$

and by (4.4) of Chapter III,

$$J_k \le C\left[k^{-1-\alpha}\|f\|^2 + \int_{x_k}^{x_{k+1}} g_1^2(t)dt\right]h^{2-\alpha} .$$

Hence $J \le J_0 + C\|f\|^2h^{2-\alpha}$. Then, in the integral J_0 , we have $u^{h'}(x) = 0$ and $|u'(x)| \le C\|f\|x^{\frac{1}{2}-\alpha}$, so

$$J_0 \le C\|f\|^2 \int_0^h x^{1-\alpha}dx = C\|f\|^2h^{2-\alpha} .$$

Finally, $J \le C\|f\|^2h^{2-\alpha}$. The positive definiteness of A_0 implies

$$\int_0^1 q(x)\left(u-u^h\right)^2 dx \le C\|f\|^2h^{2-\alpha} .$$

So, the order of the approximation is

$$\|u-u^h\|_A \le C\|f\|h^{1-\alpha/2} . \tag{1}$$

We now show that the estimate (1) is almost optimal in the sense indicated below. Consider the function $u_0(x) = (1-\gamma)^{-1}(x^{1-\gamma}-1)$, $\gamma = \text{const.} > 0$. It satisfies the boundary condition $u_0(1) = 0$ and the equation $-\left(x^{\alpha}u_0'\right)' = f_0(x)$, where $f_0(x) = -(\alpha-\gamma)x^{\alpha-\gamma-1} \in L_2(0, 1)$, if $\gamma < \alpha - \frac{1}{2}$. It is easy to check that the corresponding energy integral converges under the same condition. Let u_0^h be a piecewise linear approximation such that the abscissae of its vertices are the grid-points $x_k = kh$. We derive an estimate for J_0 . On the interval $(0, h)$, we have $\left[u_0^h(x)\right]' = \beta = \text{const.}$ The calculation of the integral J_0 is straightforward and it is easy to check that its minimum value is $Ch^{\alpha-2\gamma+1}$, $C = \text{const.}$ The minimum is attained when $\beta = (\alpha+1)h^{-\gamma}/(\alpha-\gamma+1)$. Hence, for arbitrary piecewise linear approximations $u_0^h(x)$, we have $\left\|u_0-u_0^h\right\|_A \ge \sqrt{C}\, h^{(\alpha+1)/2-\gamma}$. For arbitrary small positive ε , we put $\gamma = \alpha - \frac{1}{2} - \varepsilon$. It follows that

$$\left\|u_0-u_0^h\right\|_A \ge \sqrt{C}\, h^{1-\alpha/2-\varepsilon} . \tag{2}$$

Hence, up to the choice of ε , the estimate (1) is optimal.

A lower bound which is closed to (1) is given by the function

$$u_1(x) = \frac{x^{3/2-\alpha}}{\ln^{1+\varepsilon}(2/x)}\,\frac{1}{\ln^{1+\varepsilon}2} , \quad \varepsilon > 0 .$$

In this case, an arbitrary piecewise linear approximation u_1^h (with the abscissae of its vertices located at the grid-points $x_k = kh$, $k = 0, 1, \ldots, 2n$) satisfies the inequality $(\alpha \ne 3/2)$

$$\left\|u_1-u_1^h\right\|_A \ge \frac{\text{const}\cdot h^{1-(\alpha/2)}}{\ln^{1+\varepsilon}(2/n)} . \tag{3}$$

If $\alpha = 3/2$, then ε should be replaced by $1 + \varepsilon$.

We now prove the inequality (3). It is easy to check that $u_1 \in H_A$ and $-\left(x^{\alpha}u_1'\right)' \in L_2(0, 1)$. If u_1^h is an arbitrary piecewise linear approximation with abscissae of its vertices located at $x_k = kh$, then $\left[u_1^h(x)\right]' = \beta = \text{const}$, when $x \in (0, h)$. Consider the integral

$$J_0 = \int_0^h x^\alpha (u_1'-\beta)^2 dx = a\beta^2 - 2b\beta + c \ ,$$

where

$$a = \int_0^h x^\alpha dx = \frac{h^{1+\alpha}}{1+\alpha} \ ,$$

$$b = \int_0^h x^\alpha u_1' dx = ((3/2)-\alpha) \int_0^h \frac{\sqrt{x}}{\ln^{1+\varepsilon}(2/x)} + \dots \ ,$$

$$c = \int_0^h x^\alpha u_1'^2 dx = ((3/2)-\alpha)^2 \int_0^h \frac{x^{1-\alpha}}{\ln^{2+2\varepsilon}(2/x)} dx + \dots \ .$$

Here and below a sequence of dots denote the neglected lower order terms. The minimum value of the integral J_0 is $c - b^2/a$. This value will now be estimated. Setting $t = \ln(2/x)$ and integrating by parts, we obtain

$$b = ((3/2)-\alpha)\ \frac{2}{3}\ \frac{h^{3/2}}{\ln^{1+\varepsilon}(2/n)} + \dots \ , \qquad c = ((3/2)-\alpha)^2\ \frac{1}{2-\alpha}\ \frac{h^{2-\alpha}}{\ln^{2+2\varepsilon}(2/n)} \ .$$

Hence

$$c - b^2/a = ((3/2)-\alpha)^2 \left[\frac{1}{2-\alpha} - \frac{4(1+\alpha)}{9}\right] \frac{h^{2-\alpha}}{\ln^{2+2\varepsilon}(2/n)} + \dots \ .$$

The value in square brackets is an increasing function of α , for $1 \le \alpha \le 2$. For $\alpha = 1$, it has a minimum value which is positive. Hence it follows that, for sufficiently small h ,

$$J_0 \ge c_0 \frac{h^{2-\alpha}}{\ln^{2+2\varepsilon}(2/n)} \ , \qquad c_0 = \text{const.}$$

The estimate (3) is now an immediate consequence.

§5. Properties of the Solution

THEOREM 1. *If in the problem* (1.1), (1.8) *the value of* α *lies in the range* $1 \le \alpha \le 2$ *and* $f \in L_r(0, 1)$, $2 < r < \infty$, *then* $u \in L_r(0, 1)$.

Proof. For the proof we use the method which appears to have been developed first by Cimmino [1]. Take an arbitrary $N > 0$ and consider the function

$$\mu_N(u) = \begin{cases} |u|^{r-1} \operatorname{sign} u \ , & |u| \le N \ , \\ N^{r-1} \operatorname{sign} u \ , & |u| > N \ . \end{cases}$$

We note that $\mu_N(-u) = -\mu_N(u)$ and

$$|\mu_N(u)| = \begin{cases} |u|^{r-1}, & |u| \le N, \\ N^{r-1}, & |u| > N. \end{cases}$$

We multiply both sides of (1.1) (where $\varphi(x)$ has the form (1.4)) by $\mu_N(u)$, integrate over the interval $(0, 1)$, and evaluate the first integral by parts

$$-\mu_N(u)x^\alpha p(x)\left.\frac{du}{dx}\right|_0^1 + \int_0^1 x^\alpha p(x)u'(x)\frac{d\mu_N(u)}{dx}dx + \int_0^1 q(x)u\mu_N(u)dx = \int_0^1 f(x)\mu_N(u)dx .$$

The non-integral terms vanish. In fact, we have $u(1) = 0$, hence $\mu_N(u) = 0$. Repeating tho argument of §2 with minor changes, we obtain $\lim_{x\to 0} x^\alpha u'(x) = 0$, and the non-integral terms vanish when $x = 0$. Now

$$\int_0^1 x^\alpha p(x)u'(x)\frac{d\mu_N(u)}{dx}dx + \int_0^1 q(x)u\mu_N(u)dx = \int_0^1 f(x)\mu_N(u)dx . \tag{1}$$

We introduce the notation

$$u_N(x) = \begin{cases} u(x), & |u(x)| \le N, \\ N \operatorname{sign} u(x), & |u(x)| > N. \end{cases}$$

Consequently $|\mu_N(u)| = |u_N(x)|^{r-1}$. Consider the function under the first integral sign in (1). From the identity $|u|' = u' \cdot \operatorname{sign} u$, it follows that

$$\frac{d}{dx}\mu_N(u) = \begin{cases} (r-1)|u|^{r-2}u', & |u| < N, \\ 0, & |u| > N. \end{cases}$$

Applying this and the equality $u'u_N' = (u_N')^2$ to (1), we obtain

$$(r-1)\int_0^1 x^\alpha p(x)|u_N|^{r-2}(u_N')^2dx + \int_0^1 q(x)u\mu_N(u)dx = \int_0^1 f(x)\mu_N(u)dx . \tag{2}$$

The second integral on the l.h.s. is non-negative. Eliminating it, we obtain

$$(r-1)\int_0^1 x^\alpha p(x)|u_N|^{r-2}(u_N')^2dx \le \int_0^1 f(x)\mu_N(u)dx$$

$$\le \|f\|_r\left\{\int_0^1 |\mu_N(u)|^{r'}dx\right\}^{1/r'} = \|f\|_r \cdot \|u_N\|_r^{r-1} . \tag{3}$$

Now

$$\frac{d}{dx}|u_N|^{r/2} = \frac{r}{2}|u_N|^{(r/2)-1}u_N' \operatorname{sign} u_N$$

so the inequality (3) takes the form

$$\frac{4(r-1)}{r^2}\int_0^1 x^\alpha p(x)\left[\frac{d}{dx}|u_N|^{r/2}\right]^2 dx \le \|f\|_r \cdot \|u_N\|_r^{r-1} .$$

Since the operator A_0 is positive definite and, obviously, $|u_N|^{r/2} \in H_{A_0}$, it follows that

$$\int_0^1 x^\alpha p(x)\left[\frac{d}{dx}|u_N|^{r/2}\right]^2 dx \ge \varkappa^2 \int_0^1 |u_N|^r dx = \varkappa^2\|u_N\|_r^r , \quad \varkappa = \text{const.} > 0 .$$

Now we obtain from (3) that

$$\|u_N\|_r \le \frac{r^2}{4(r-1)\varkappa^2}\|f\|_r .$$

With $N \to \infty$ this yields

$$\|u\|_r \le \frac{r^2}{4(r-1)\varkappa^2}\|f\|_r , \tag{4}$$

from which the required result follows.

THEOREM 2. *Let* $q(x) \ge q_0 = \text{const.} > 0$. *If* $\alpha \ge 1$ *and* $f \in L_r(0, 1)$, $2 < r \le \infty$, *then* $u \in L_r(0, 1)$.

Proof. For any $\alpha \ge 1$, $\lim_{x\to 0} x^\alpha u'(x) = 0$, so for any $\alpha \ge 1$ the relations (1) and (2) hold. We eliminate the first term on the l.h.s. of (2) and replace $q(x)$ by q_0 in the second term. This leads to

$$q_0 \int_0^1 u\mu_N(u)dx \le \int_0^1 f\mu_N(u)dx \le \|f\|_r \cdot \|u_N\|_r^{r-1} .$$

Since $u\cdot\mu_N(u) = |u|\cdot|\mu_N(u)| \ge |u_N|\cdot|\mu_N(u)| = |u_N|^r$, it follows that $\|u_N\|_r \le q_0^{-1}\|f\|_r$. In the limit as $r \to \infty$, this yields

$$\|u\|_r \le q_0^{-1}\|f\|_r . \tag{5}$$

If f is a bounded function, we can let $r \to \infty$ in (5) yielding

$$\|u\|_\infty \le q_0^{-1}\|f\|_\infty . \tag{5_1}$$

§6. Improved Estimates

We again consider the problem (1.1), (1.8) for $1 \le \alpha \le 2$. As previously, we set $g(x) = f(x) - q(x)u(x)$ and let $f \in L_r(0, 1)$, $2 < r < \infty$. Then $g \in L_r(0, 1)$ and $\|g\|_r \le c\|f\|_r$. Integrating both sides of (1.1) w.r.t. x, we obtain (3.2) with $\gamma = 0$:

$$-x^{\alpha}p(x)\,\frac{du}{dx} = \int_0^x g(t)dt\ . \tag{1}$$

By Hölder inequality, we obtain

$$\left|x^{\alpha}p(x)\,\frac{du}{dx}\right| \le x^{1-(1/r)}\left\{\int_0^x |g(t)|^r dt\right\}^{1/r} \le Cx^{1-(1/r)}\|f\|_r$$

and hence

$$\left|\frac{du}{dx}\right| \le Cx^{1-(1/r)-\alpha}\|f\|_r\ . \tag{2}$$

Then,

$$\frac{d^2u}{dx^2} = -\frac{(x^{\alpha}p)'}{x^{\alpha}p}\,\frac{du}{dx} + \frac{g(x)}{x^{\alpha}p} = O\big(x^{-(1/r)-\alpha}\big)\|f\|_r + x^{-\alpha}g_1(x)\ ,$$

where $g_1 \in L_r(0, 1)$ and $\|g_1\|_r \le C\|f\|_r$. With $k \ge 1$, we estimate the integral

$$\int_{x_k}^{x_{k+1}} [u''(t)]^2dt \le C\|f\|_r^2 \int_{x_k}^{x_{k+1}} t^{-(2/r)-2\alpha}dt + 2\int_{x_k}^{x_{k+1}} t^{-2\alpha}g_1^2(t)dt\ .$$

Obviously

$$\int_{x_k}^{x_{k+1}} t^{-(2/r)-2\alpha}dt \le \frac{h^{1-(2/r)-2\alpha}}{k^{(2/r)+2\alpha}}\ ,$$

$$\int_{x_k}^{x_{k+1}} t^{-2\alpha}g_1^2(t)dt \le \frac{h^{-2\alpha}}{k^{2\alpha}}\int_{x_k}^{x_{k+1}} g_1^2(t)dt\ .$$

From the above inequalities it follows that

$$\int_{x_k}^{x_{k+1}} [u''(t)]^2dt \le Ch^{1-(2/r)-2\alpha}\left[k^{-(2/r)-2\alpha}\|f\|_r^2 + k^{-2\alpha}\left\{\int_{x_k}^{x_{k+1}} |g_1(t)|^r dt\right\}^{2/r}\right]\ .$$

We take the same piecewise linear approximation as in §4, and thereby obtain

$$J_k = \int_{x_k}^{x_{k+1}} x^\alpha p(x)|u'(x)-u^{h\prime}(x)|^2 dx \le (k+1)^\alpha h^\alpha p_1 \int_{x_k}^{x_{k+1}} |u'(x)-u^{h\prime}(x)|^2 dx$$

$$\le 2^\alpha k^\alpha h^{2+\alpha} p_1 \int_{x_k}^{x_{k+1}} [u''(t)]^2 dt \le Ch^{3-(2/r)-\alpha}\left[k^{-(2/r)-\alpha}\|f\|_r^2 + k^{-\alpha}\left\{\int_{x_k}^{x_{k+1}} |g_1(t)|^r dt\right\}^{2/r}\right].$$

Summing and taking into account that the series $\sum_{k=1}^{\infty} k^{-(2/r)-\alpha}$ converges, we obtain

$$J - J_0 = \sum_{k=1}^{2n-1} J_k \le Ch^{3-(2/r)-\alpha}\left[\|f\|_r^2 + \sum_{k=1}^{2n-1} k^{-\alpha}\left\{\int_{x_k}^{x_{k+1}} |g_1(t)|^r\right\}^{2/r}\right].$$

We estimate the last sum using Hölder inequality with exponents $r/(r-2)$ and $r/2$:

$$\sum_{k=1}^{2n-1} k^{-\alpha}\left\{\int_{x_k}^{x_{k+1}} |g_1(t)|^r dt\right\}^{2/r} \le \left\{\sum_{k=1}^{2n-1} k^{-(\alpha r)/(r-2)}\right\}^{(r-2)/r} \left\{\int_h^1 |g_1(t)|^r dt\right\}^{2/r}.$$

Since $r > 2$ and $\alpha \ge 1$, the series $\sum_{k=1}^{\infty} k^{-(\alpha r)/(r-2)}$ converges and the r.h.s. is dominated by $C\|g_1\|_r^2 \le C\|f\|_r^2$. Consequently,

$$J - J_0 \le Ch^{3-(2/r)-\alpha}\|f\|_r^2 .$$

The integral J_0 is estimated by using the inequality (2):

$$J_0 = \int_0^h x^\alpha p(x) u'^2 dx \le Ch^{3-(2/r)-\alpha}\|f\|_r^2 .$$

Hence $J \le Ch^{3-(2/r)-\alpha}\|f\|_r^2$. The same estimate is valid for the integral

$$\int_0^1 q\left(u-u^h\right)^2 dx$$

since A_0 is positive-definite. Consequently,

$$\|u-u^h\|_A \le C\|f\|_r h^{((3-\alpha)/2)-(1/r)} . \tag{3}$$

We now assume that $r = \infty$ and condition (T) of §1 holds. Then $g(x)$ is a bounded function. The subsequent argument is the same as above. It is only necessary to set $r = \infty$ to obtain

$$\|u-u^h\|_A \le C\|f\|_\infty h^{(3-\alpha)/2} . \tag{4}$$

When $\alpha = 1$, the order is $O(h)$ which corresponds to that for the non-degenerate case.

§7. The case $\alpha \geq 2$

The following results hold only for $\alpha < (5/2) - (1/r)$. As before, we have

$$\int_0^1 x^\alpha p(x)|u'(x)-u^{h\prime}(x)|^2 dx \leq C\|f\|_r^2 h^{3-\alpha-(2/r)} \quad . \tag{1}$$

Consequently,

$$\int_0^1 q(x)\left(u-u^h\right)^2 dx \leq q_1 \int_0^1 \left(u-u^h\right)^2 dx \stackrel{\text{def}}{=\!=} q_1 \mathcal{D} \ , \quad q_1 = \sup q(x) \ . \tag{2}$$

We introduce the notation

$$\mathcal{D}_k = \int_{x_k}^{x_{k+1}} \left(u-u^h\right)^2 dx \ , \quad \sum_{k=0}^{2n-1} \mathcal{D}_k = \mathcal{D} \ .$$

Applying (6.2) along with (4.3) of Chapter III, we find

$$\sum_{k=1}^{2n-1} \mathcal{D}_k \leq C\|f\|_r^2 h^{5-(2/r)-2\alpha} \sum_{k=1}^{2n-1} k^{-\left(2\alpha+(2/r)-2\right)} \leq C\|f\|_r^2 h^{5-(2/r)-2\alpha} \quad . \tag{3}$$

The estimate (3) is meaningful only if

$$\alpha < (5/2) - (1/r) \ . \tag{4}$$

It remains to give an estimate for

$$\mathcal{D}_0 = \int_0^h [u(x)-u(h)]^2 dx \leq 2\int_0^h u^2(x)dx + 2hu^2(h) \ .$$

Integrating (6.2) over the range from x to 1 , we find that $|u(x)| \leq C\|f\|_r x^{2-\alpha-(1/r)}$. Hence $\mathcal{D}_0 \leq C\|f\|_r^2 h^{5-2\alpha-(2/r)}$. Along with (3) this gives

$$\int_0^1 q\left(u-u^h\right)^2 dx \leq C\|f\|_r^2 h^{5-2\alpha-(2/r)} \quad . \tag{5}$$

We sum the inequalities (1) and (5). Since $3-\alpha \geq 5-2\alpha$, we finally obtain

$$\|u-u^h\|_A \leq C\|f\|_r h^{(5/2)-\alpha-(2/r)} \quad . \tag{6}$$

NOTE 1. We set $x^\gamma = \tau$, $\gamma = \text{const.} > 1$. Equation (1.1) takes the form (3.8), where this time

$$P(\tau) = \gamma p(x)\tau^{(\alpha+\gamma-1)/\gamma} \ , \quad Q(\tau) = \frac{q(x)}{\gamma\tau^{(\gamma-1)/\gamma}} \ , \quad F(\tau) = \frac{f(x)}{\gamma\tau^{(\gamma-1)/\gamma}} \ .$$

The new degeneracy exponent condition is $(\alpha+\gamma-1)/\gamma < \alpha$. If there exists a $\gamma > 1$

such that $Q(\tau)$ is a bounded function and $F \in L_r(0, 1)$, $r \geq 2$, then the estimates for the approximation can be improved.

NOTE 2. Assume that $\alpha \geq 2$, $2 < r \leq \infty$, and condition (T) of §1 holds. From (6) and the Theorem of §7, Chapter III, it follows that A^{-1} is completely continuous and maps $L_r(0, 1)$ into H_A , if α and r satisfy (4). When $\alpha = 2$ and $r < \infty$, the above condition on $q(x)$ is not necessary. It is sufficient that $q(x) \geq 0$.

§8. More General Equations

We again consider equation (1.1) but $\varphi(x)$ does not now have the form (1.4). We introduce the following terminology. Let $\psi(x)$ be continuous and be of one sign on the interval $(0, a)$. Assume that there exist a scalar α , $-\infty \leq \alpha \leq \infty$, such that

$$\lim_{x \to 0} x^{-\sigma}\psi(x) = 0 , \quad \sigma < \alpha , \quad \lim_{x \to 0} x^{-\sigma}\psi(x) = \infty , \quad \sigma > \alpha .$$

The scalar α will be called the *degeneracy exponent* of the function $\psi(x)$ at the point $x = 0$. The qualifier "at the point $x = 0$ " will be omitted below.

We shall consider equations of the form (1.1) in which the coefficient $\varphi(x)$ has a degeneracy exponent $\alpha > 0$. If $\alpha < 1$, then the integral (1.2) converges. If $1 < \alpha < 2$, then the integral (1.2) diverges and the integral (1.3) converges. If $\alpha > 2$, then the integral (1.3) diverges.

Let the derivative $\varphi'(x)$ have a degeneracy exponent $\alpha - 1$. *If* α *is from one of the intervals* (0, 1), (1, 2) *or* $\left(2, (5/2)-(1/r)\right)$ *, then the estimates of the preceeding sections hold. It is only necessary to replace the power* α *of* h *by* $\alpha + \varepsilon$ *, where* ε *is an arbitrary small positive number.* Below we give a proof.

(a) Assume that $0 < \alpha < 1$. The operator A_0 is positive definite, the solution of the problem (1.1), (1.7) exists, belongs to the space $L_2(0, 1)$ and $g = (f-qu) \in L_2(0, 1)$, $\|g\| \leq C\|f\|$. Integrating, we find

$$u'(x) = \frac{1}{\varphi(x)} \int_0^x g(t)dt + \frac{C}{\varphi(x)} .$$

The principal term is the second, so $|u'(x)| \leq C\|f\|/\varphi(x)$. Then,

$$u''(x) = -\frac{\varphi'(x)}{\varphi(x)} u'(x) - \frac{g(x)}{\varphi(x)} .$$

Let τ be an arbitrary positive number. Then

$$\lim_{x \to 0} x^{-\alpha+\tau}\varphi(x) = 0 , \quad \lim_{x \to 0} x^{-\alpha-\tau}\varphi(x) = \infty .$$

Hence

$$C_1 x^{\alpha+\tau} \le \varphi(x) \le C_2 x^{\alpha-\tau} \ , \quad C_1,\ C_2 = \text{const} \ . \tag{1}$$

Analogously,

$$C_3 x^{\alpha+\tau-1} \le \varphi'(x) \le C_4 x^{\alpha-\tau-1} \ , \quad C_3,\ C_4 = \text{const} \ . \tag{2}$$

But

$$\left| \frac{\varphi'(x)}{\varphi(x)}\, u'(x) \right| \le C\|f\| \, \frac{|\varphi'(x)|}{\varphi^2(x)} \le C\|f\| x^{-\alpha-3\tau-1} \ .$$

This gives

$$|u''(x)| \le C\|f\| x^{-\alpha-3\tau-1} + C|g(x)|\, x^{-\alpha-\tau} \ . \tag{3}$$

In the same way as in the preceeding sections, we set

$$J = \int_0^1 \varphi(x)\, |u'(x) - u^{h\prime}(x)|^2 dx \ ,$$

$$J_k = \int_{x_k}^{x_{k+1}} \varphi(x)\, |u'(x) - u^{h\prime}(x)|^2 dx \ .$$

If $k \ge 1$, then by (4.4) of Chapter III,

$$J_k \le C x_{k+1}^{\alpha-\tau} \int_{x_k}^{x_{k+1}} |u'(x) - u^{h\prime}(x)|^2 dx \le C(k+1)^{\alpha-\tau} h^{2+\alpha-\tau} \int_{x_k}^{x_{k+1}} [u''(t)]^2 dt$$

$$\le C(k+1)^{\alpha-\tau} h^{2+\alpha-\tau} \left[\|f\|^2 \int_{x_k}^{x_{k+1}} t^{-2-2\alpha-6\tau} dt + \int_{x_k}^{x_{k+1}} g^2(t)\, t^{-2\alpha-2\tau} dt \right] .$$

Hence

$$J_k \le C(k+1)^{\alpha-\tau} h^{2+\alpha-\tau} \left[\|f\|^2 k^{-2-2\alpha-6\tau} h^{-1-2\alpha-6\tau} + k^{-2\alpha-2\tau} h^{-2\alpha-2\tau} \int_{x_k}^{x_{k+1}} g^2(t) dt \right]$$

$$\le C\|f\|^2 k^{-2-\alpha-7\tau} h^{1-\alpha-7\tau} + C k^{-\alpha-3\tau} h^{2-\alpha-3\tau} \int_{x_k}^{x_{k+1}} g^2(t) dt$$

$$\le \frac{C\|f\|^2}{k^{2+\alpha}}\, h^{1-\alpha-7\tau} + C h^{2-\alpha-3\tau} \int_{x_k}^{x_{k+1}} \bar{g}^2(t) dt \ .$$

Summing, we obtain

$$J - J_0 \le C\|f\|^2 h^{1-\alpha-7\tau} \ . \tag{4}$$

The estimation of J_0 is straightforward

$$J_0 = \int_0^h \varphi(x)\left[u'(x) - \frac{u(h)}{h}\right]^2 dx \le 2\int_0^h \varphi(x)u'^2(x)dx + 2\,\frac{u^2(h)}{h^2}\int_0^h \varphi(x)dx\ . \tag{5}$$

The first integral on the r.h.s. is dominated by

$$C\|f\|^2 \int_0^h \frac{dx}{\varphi(x)} \le C\|f\|^2 \int_0^h x^{-\alpha-\tau}dx = C\|f\|^2 h^{1-\alpha-\tau}\ .$$

We now estimate $u(h)$. We have

$$|u(x)| = \left|\int_0^x u'(t)dt\right| \le C\|f\| \int_0^x \frac{dx}{\varphi(x)} \le C\|f\|x^{1-\alpha-\tau}\ .$$

Clearly, the second term in (5) is dominated by

$$C\|f\|^2 h^{-2\alpha-2\tau}\int_0^h \varphi(x)dx = C\|f\|^2 h^{1-\alpha-3\tau}\ .$$

Consequently, $J_0 \le ch^{1-\alpha-7\tau}\|f\|^2$. We combine this with inequality (4), write $7\tau = \varepsilon$, and thereby obtain $J \le c\|f\|^2 h^{1-\alpha-\varepsilon}$. The same estimate holds for the integral

$$\int_0^1 q\left(u-u^h\right)^2 dx$$

since A_0 is a positive definite operator. As a result we obtain the estimate

$$\|u-u^h\|_A \le C\|f\|h^{(1-\alpha-\varepsilon)/2}\ . \tag{6}$$

The estimate (6) is "optimal". In order to verify this we consider the function $u_0(x)$ which is defined on the interval $[0, \frac{1}{2}]$ by the formula

$$u_0(x) = \int_0^x \frac{dt}{\varphi(t)}\ ,$$

belongs to the class C^∞ on the interval $[\frac{1}{2}, 1]$, is continuous along with its first and second derivatives when $x = \frac{1}{2}$, and vanishes when $x = 1$. We write

$$-\frac{d}{dx}\varphi(x)\frac{du_0}{dx} = f_0(x)\ .$$

Obviously, $f_0 \in C^2[0, 1] \cap C^\infty(\frac{1}{2}, 1]$ with $f_0(x) = 0$, $x \in [0, \frac{1}{2}]$.

Let β be a constant. We evaluate the integral

$$\int_0^h \varphi(x)|u_0'(x)-\beta|^2 dx = \int_0^h \left[\frac{1}{\varphi(x)} - 2\beta + \beta^2\varphi(x)\right]dx = \int_0^h \frac{dx}{\varphi(x)} + O(h)\ .$$

From (1) we have $1/\varphi(x) \geq C_2^{-1}x^{-\alpha+\tau}$ and

$$\int_0^h \frac{dx}{\varphi(x)} \geq \frac{1}{C_2}\int_0^h x^{-\alpha+\tau}dx = \frac{1}{C_2(1-\alpha+\tau)}\, h^{1-\alpha+\tau} ,$$

so, if h is sufficiently small, then

$$\int_0^h \varphi(x)\,|u_0'(x)-\beta|^2 dx \geq Ch^{1-\alpha+\tau} .$$

Thus, for any piecewise linear approximation u_0^h with the abscissae of the vertices located at kh , $k = 1, 2, \ldots, 2n$, we obtain

$$\|u_0-u_0^h\|_A \geq Ch^{(1-\alpha+\tau)/2} .$$

(b) Assume that $1 < \alpha < 2$, $f \in L_r(0, 1)$, and $2 \leq r \leq \infty$. If $r = \infty$, then we assume that condition (T) of §1 holds. Under these assumptions, $u \in L_r(0, 1)$. For $r = 2$, it follows from the positive definiteness of the operator A , and for $r > 2$ it can be proved by the Cimmino method. Now

$$g = (f-qu) \in L_r(0, 1) , \quad \|g\|_r \leq C\|f\|_r ,$$

$$\varphi(x)u'(x) = -\int_0^x g(t)dt = O\left(x^{1-(1/r)}\right)\|f\|_r ,$$

and hence

$$|u'(x)| \leq C\|f\|_r x^{1-\alpha-\tau-(1/r)} . \tag{7}$$

In this way, we obtain an estimate for the second derivative

$$[u''(x)]^2 = \left[\frac{\varphi'(x)}{\varphi(x)}\,u'(x) + \frac{g(x)}{\varphi(x)}\right]^2 \leq C\|f\|_r^2 x^{-2\alpha-6\tau-(2/r)} + Cg^2(x)x^{-2\alpha-2\tau} .$$

For the integral J_k , $k \geq 1$, we have

$$J_k = \int_{x_k}^{x_{k+1}} \varphi(x)|u'(x)-u^{h'}(x)|^2 dx$$

$$\le Cx_{k+1}^{\alpha-\tau} \int_{x_k}^{x_{k+1}} |u'(x)-u^{h'}(x)|^2 dx \le Ck^{\alpha-\tau}h^{2+\alpha-\tau} \int_{x_k}^{x_{k+1}} [u''(x)]^2 dx$$

$$\le Ck^{\alpha-\tau}h^{2+\alpha-\tau}\left[\|f\|_r^2 \int_{x_k}^{x_{k+1}} x^{-2\alpha-6\tau-(2/r)}dx + \int_{x_k}^{x_{k+1}} x^{-2\alpha-2\tau}g^2(x)dx\right]$$

$$\le Ck^{\alpha-\tau}h^{2+\alpha-\tau}\left[k^{-2\alpha-6\tau-(2/r)}h^{1-2\alpha-6\tau-(2/r)}\|f\|_r^2 + k^{-2\alpha-2\tau}\int_{x_k}^{x_{k+1}} g^2(x)dx\right] . \quad (8)$$

Applying the Hölder inequality in the same way as in §6, we obtain

$$\int_{x_k}^{x_{k+1}} g^2(x)dx \le Ch^{1-(2/r)}\|f\|_r^2 .$$

Consequently, we see that the first term on the r.h.s. of (8) is the principal term, and therefore

$$J_k \le Ck^{-\alpha-7\tau-(2/r)}h^{3-\alpha-7\tau-(2/r)}\|f\|_r^2$$

and

$$J - J_0 \le Ch^{3-\alpha-\varepsilon-(2/r)}\|f\|_r^2 , \quad \varepsilon = 7\tau . \quad (9)$$

Further

$$J_0 = \int_0^h \varphi(x)u'^2 dx \le C\|f\|_r^2 \int_0^h x^{2-\alpha-3\tau-(2/r)}dx \le C\|f\|_r^2 h^{3-\alpha-\varepsilon-(2/r)} .$$

Combining the last with (9), we derive the estimate

$$J \le C\|f\|_r^2 h^{3-\alpha-\varepsilon-(2/r)} . \quad (10)$$

Since A_0 is positive definite for $\alpha > 2$, we have

$$\int_0^1 q(x)|u(x)-u^h(x)|^2 dx \le CJ \quad (11)$$

and consequently

$$\|u-u^h\|_A \le C\|f\|_r h^{((3-\alpha-\varepsilon)/2)-(1/r)} . \quad (12)$$

(c) Assume that $\alpha > 2$. In this case, we must derive an independent estimate for the integral (11). As in §7, we arrive at (7.2). Then, by (7), we have for

$k \geq 1$,

$$E_k \leq C\|f\|_r^2 h^2 \int_{x_k}^{x_{k+1}} x^{2-2\alpha-2\tau-(2/r)} dx \leq C\|f\|_r k^{3-2\alpha-2\tau-(2/r)} h^{5-2\alpha-2\tau-(2/r)}$$

and hence

$$\sum_{k=1}^{2n-1} E_k \leq C\|f\|_r^2 h^{5-2\alpha-2\tau-(2/r)} . \tag{13}$$

Using the same arguments as in §7, we find that E_0 has an estimate of the form (12). From this the final estimate easily follows

$$\|u-u^h\|_{\mathsf{A}} \leq C\|f\|_r h^{5-2\alpha-\varepsilon-(1/r)} . \tag{14}$$

This estimate has sense for $2 < \alpha < (5/2) - (1/r)$.

§9. Approximations in L_2

Now we derive an estimate for $\|u-u^h\| = \|u-u^h\|_{L_2}$, where $u^h(x)$ is the piecewise linear approximation considered in the preceeding sections of this chapter. For simplicity of presentation we limit consideration to the case when the coefficient $\varphi(x)$ of (1.1) has the form (4.1). We write

$$\mathcal{D} = \|u-u^h\|^2 = \int_0^1 |u(x)-u^h(x)|^2 dx$$

and

$$\mathcal{D}_k = \int_{x_k}^{x_{k+1}} |u-u^h(x)|^2 dx ,$$

so that

$$\mathcal{D} = \sum_{k=0}^{2n-1} \mathcal{D}_k .$$

Applying Friedrichs' inequality

$$\mathcal{D}_k \leq h^2 \int_{x_k}^{x_{k+1}} |u'(x)-u^{h\prime}(x)|^2 dx \leq \frac{h^{2-\alpha}}{p_0 k^\alpha} \int_{x_k}^{x_{k+1}} x^\alpha p(x)|u'(x)-u^{h\prime}(x)|^2 dx$$

$$\leq \frac{h^{2-\alpha}}{p_0} \int_{x_k}^{x_{k+1}} x^\alpha p(x)|u'(x)-u^{h\prime}(x)|^2 dx .$$

Summing, we obtain

$$\mathcal{D} \le \mathcal{D}_0 + \frac{h^{2-\alpha}}{p_0}\int_h^1 x^\alpha p(x)|u'(x)-u^{h'}(x)|^2 dx \ . \tag{1}$$

We now estimate $\mathcal{D}_0$. Assuming that $0 < \alpha < 2$, A_0 is positive definite (see §1) and $(\gamma = \text{const} > 0)$

$$\int_0^h u^2(x)dx \le \gamma\int_0^1 x^\alpha p(x)u'^2(x)dx \le \gamma p_1\int_0^1 x^\alpha u'^2(x)dx \ , \quad \forall u \in H_{A_0} \ .$$

Setting $t = hx$, $v(t) = u(h^{-1}t)$, we obtain

$$\int_0^h v^2 dt \le \gamma p_1 h^{2-\alpha}\int_0^h t^\alpha[v'(t)]^2 dt \le \frac{\gamma p_1}{p_0} h^{2-\alpha}\int_0^h t^\alpha p(t)[v'(t)]^2 dt \ .$$

Replacing $v(t)$ by $u(t) - u^h(t)$, we find

$$\mathcal{D}_0 = \int_0^h |u(t)-u^h(t)|^2 dt \le \frac{\gamma p_1}{p_1} h^{2-\alpha}\int_0^h t^\alpha p(t)|u'(t)-u^{h'}(t)|^2 dt \ . \tag{2}$$

Summing (1) and (2), we arrive at

$$\|u-u^h\|^2 \le Ch^{2-\alpha}\int_0^1 x^\alpha p(x)|u'(x)-u^{h'}(x)|^2 dx \le Ch^{2-\alpha}\|u-u^h\|^2 \ . \tag{3}$$

Using (3.7), (4.1) and (6.3) we derive the estimates

$$\|u-u^h\| \le Ch^{(3/2)-\alpha}\|f\| \ , \quad 0 < \alpha < 1 \ , \tag{4}$$

$$\|u-u^h\| \le Ch^{(5/2)-\alpha-(1/r)}\|f\| \ , \quad 1 \le \alpha < 2 \ , \quad 2 < r < \infty \ . \tag{5}$$

The estimate (5) also holds for $r = \infty$, if condition (T) of §1 holds.

For the case $2 \le \alpha < (5/2) - (1/r)$ with $2 < r \le \infty$, the estimate follows from (7.5). It has the same form as (5).

§10. Other Boundary Conditions

We shall consider briefly the cases when the boundary conditions do not have the form (1.7) or (1.8). For example let the condition $u(1) = 0$ be replaced by

$$u'(1) + au(1) = 0 \ , \quad a > 0 \tag{1}$$

and the condition $u(0) = 0$ by

$$\lim_{x\to 0} x^\alpha u'(x) = 0 \ . \tag{2}$$

Condition (2) applies only when $0 < \alpha < 1$. It is sufficient to modify the piecewise linear approximation $u^h(x)$ (approximating the solution $u(x)$) in the following way: if $0 < \alpha < 1$ and conditions (1), (2) hold, we take as the vertices of the piecewise approximation the points $(x_k, u(x_k))$, $x_k = kh$, $k = 0, 1, \ldots, 2n$; if $\alpha \geq 1$ and only condition (1) holds, then the vertices should be

$$(0, u(h)), (h, u(h)), (2h, u(2h)), \ldots, (1, u(1)) .$$

In this way, the arguments of §§3-9 go through without any substantial changes, but the estimates themselves must be changed slightly.

CHAPTER VI

SOME DEGENERATE TWO-DIMENSIONAL NORMS

In this chapter we consider two kinds of two-dimensional norms: one is generated by the Dirichlet integral written with respect to polar coordinates, the other is an energy norm generated by certain second order degenerate differential equations. The former is connected naturally with radially symmetric grids (see Mikhlin [13]), which are widely used for problems arising in technology. An important advantage of radially-symmetric grids is that they are perfect for problems involving a radial or axial symmetric structure. Below, we show that radially symmetric grids on circles give practically the same order of approximation as a rectangular grid on rectangulars. To date, insufficient attention has been paid to norms appropriate for degenerate equations. We confine attention to the results of Gusman and Oganesjan [1] (see §6 below).

§1. Approximations for Radially-Symmetric Grids

Let κ denote the circle $x_1^2 + x_2^2 \leq 1$, cut along the radius $x_2 = 0$, $0 < x_1 \leq 1$, or some part of it. We introduce the polar coordinates r and ϑ with the same origin as the Cartesian coordinates, and construct the radially symmetric grid (see Figure 16) with step size h_1 along r and h_2 w.r.t. ϑ .

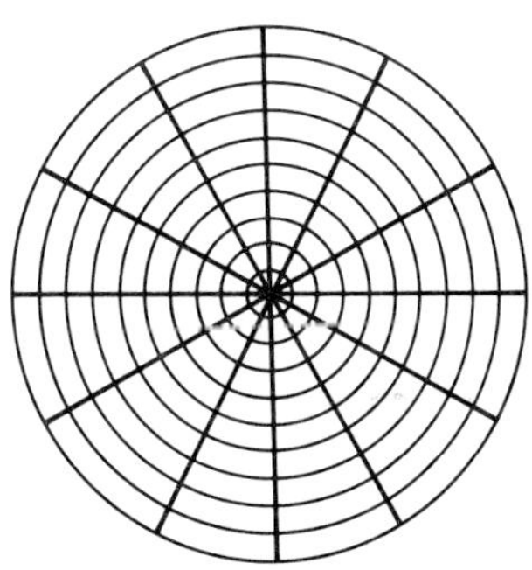

Fig. 16

We take $h_1 = 1/2n_1$, $h_2 = \pi/n_2$, where n_1, n_2 are integers. The circle κ will be completely covered by the curved rectangles of the gird. Let

$$\omega(t) = \begin{cases} t , & 0 \le t \le 1 , \\ 2-t , & 1 < t \le 2 , \\ 0 , & t \notin [0, 2] , \end{cases}$$

be the one-dimensional primitive function of zero degree constructed in §6 of Chapter II. This function can be used to construct two-dimensional coordinate functions for the variational-difference method as follows

$$\varphi_{jk}^{(h)}(r, \vartheta) = \omega\left(\frac{r}{h_1} - j\right)\omega\left(\frac{\vartheta}{h_2} - k\right) . \tag{1}$$

We denote by $C_0^{(2)}(\kappa)$ the class of functions which have continuous second derivatives with respect to x_1 and x_2 inside κ , and vanish on the boundary $r = 1$. With respect to the norm of $W_2^{(1)}(\kappa)$, we derive an estimate of the order of approximation of functions from $C_0^{(2)}(\kappa)$ by linear combinations of the functions (1). We note that, for the uncut circle κ , the $W_2^{(1)}$ norm for functions of $C_0^{(2)}(\kappa)$ is equivalent to the energy norm of a first boundary value problem for a second order non-degenerate elliptic equation; the later is, in its turn, equivalent to the norm generated by the Dirichlet integral. If $u \in C_0^{(2)}(\kappa)$, then we take the following approximating function

$$u^h(r, \vartheta) = \sum_{j=-1}^{2n_1-2} \sum_{k=-1}^{2n_2-1} u\big((j+1)h_1, (k+1)h_2\big)\omega\left(\frac{r}{h_1} - j\right)\omega\left(\frac{\vartheta}{h_2} - k\right) .. \tag{2}$$

The terms corresponding to $j = -1$, $k = -1$, $k = 2n_2 - 1$ arise because the function $u(r, \vartheta)$ does not in general vanish along the sides $r = 0$, $\vartheta = 0$, $\vartheta = 2\pi$ of the rectangle $0 \le r \le 1$, $0 \le \vartheta \le 2\pi$ in the (r, ϑ) plane.

We simplify (2) by removing terms with $j = -1$. Their sum is

$$\omega\left(\frac{r}{h_1} + 1\right) \sum_{k=-1}^{2n_2-1} u\big(0, (k+1)h_2\big)\omega\left(\frac{\vartheta}{h_2} - k\right) . \tag{3}$$

When $r = 0$ the value of u does not depend on ϑ . Denoting this value by $u(0)$, we rewrite (3) as

$$u(0)\omega\left(\frac{r}{h_1} + 1\right) \sum_{k=-1}^{2n_2-1} \omega\left(\frac{\vartheta}{h_2} - k\right) . \tag{4}$$

It can be easily seen that the sum in (4) is equal to unity. In fact, on any section of the form $lh_2 < \vartheta < (l+1)h_2$, $l = -1, 0, \ldots, 2n_2-1$, the only non-vanishing terms are $\omega\left(\frac{\vartheta}{h_2} - l\right)$ and $\omega\left(\frac{\vartheta}{h_2} - l + 1\right)$. The argument of the first lies between zero and unity, the argument of the second between unity and two, therefore $\omega\left(\frac{\vartheta}{h_2} - l\right) = \frac{\vartheta}{h_2} - l$ and $\omega\left(\frac{\vartheta}{h_2} - l + 1\right) = 1 + l - \frac{\vartheta}{h_2}$ with their sum equal to unity. Thus, (4) equals $u(0)\omega\left(\frac{r}{h_1} + 1\right)$ and (2) takes the form

$$u^h(r, \vartheta) = u(0)\omega\left(\frac{r}{h_1} + 1\right) + \sum_{j=0}^{2n_1-2} \sum_{k=-1}^{2n_2-1} u\big((j+1)h_1, (k+1)h_2\big)\omega\left(\frac{r}{h_1} - j\right)\omega\left(\frac{\vartheta}{h_2} - k\right) . \quad (5)$$

We note that $\omega\left(\frac{r}{h_1} + 1\right) = 0$ outside the interval $(0, h_1)$, and $\omega\left(\frac{r}{h_1} + 1\right) = 1 - \frac{r}{h_2}$ on this interval. In addition, the values of u along the uncut part of the radius $x_2 = 0$, $0 < x_1 < 1$, coincide for $k = -1$ and $k = 2n_2 - 1$.

We derive an estimate for

$$\|u(r, \vartheta)-u^h(r, \vartheta)\|_{W_2^{(1)}(K)}$$

or, in other words, for the integral

$$\int_0^1\int_0^{2\pi}\left[r\left(\frac{\partial u}{\partial r} - \frac{\partial u^h}{\partial r}\right)^2 + \frac{1}{r}\left(\frac{\partial u}{\partial \vartheta} - \frac{\partial u^h}{\partial \vartheta}\right)^2\right]d\vartheta dr = \int_0^1\int_0^{2\pi}\left\{r\left[\frac{\partial u}{\partial r} - \frac{u(0)}{h_1}\omega'\left(\frac{r}{h_1} + 1\right)\right.\right.$$

$$\left. - \frac{1}{h_1}\sum_{j=0}^{2n_1-2}\sum_{k=-1}^{2n_2-1} u\big((j+1)h_1, (k+1)h_2\big)\omega'\left(\frac{r}{h_1} - j\right)\omega\left(\frac{\vartheta}{h_2} - k\right)\right]^2$$

$$\left. + \frac{1}{r}\left[\frac{\partial u}{\partial \vartheta} - \frac{1}{h_2}\sum_{j=0}^{2n_1-2}\sum_{k=-1}^{2n_2-1} u\big((j+1)h_1, (k+1)h_2\big)\omega\left(\frac{r}{h_1} - j\right)\omega'\left(\frac{\vartheta}{h_2} - k\right)\right]^2\right\}d\vartheta dr . \quad (6)$$

§2. Estimation of the First Integral

We estimate the integral (1.6) as the sum of integrals over the curved rectangles with sides h_1 and h_2 . We denote by $Q_{j_0k_0}$ the rectangle with lower vertex (j_0h_1, k_0h_2) and consider the corresponding integral. It can be easily shown that the terms of the sum under the integral sign which do not vanish correspond to the situations when $j_0 - j$ and $k_0 - k$ are either zero or unity. In the notation of Chapter I this can be rewritten as $(j_0-j, k_0-k) \in I$.

Consequently, we must consider the integral

$$\iint\limits_{Q_{j_0k_0}} \left\{ r\left[\frac{\partial u}{\partial r} - \frac{u(0)}{h_1}\,\omega'\left(\frac{r}{h_1}+1\right)\right.\right.$$

$$\left. - \frac{1}{h_1}\sum_{(j_0-j,k_0-k)\in I} u\big((j+1)h_1,\ (k+1)h_2\big)\omega'\left(\frac{r}{h_1}-j\right)\omega\left(\frac{\vartheta}{h_2}-k\right)\right]^2$$

$$\left. + \frac{1}{r}\left[\frac{\partial u}{\partial \vartheta} - \frac{1}{h_2}\sum_{(j_0-j,k_0-k)\in I} u\big((j+1)h_1,\ (k+1)h_2\big)\omega\left(\frac{r}{h_1}-j\right)\omega'\left(\frac{\vartheta}{h_2}-k\right)\right]^2\right\} d\vartheta dr \qquad (1)$$

which is the sum of two integrals. We initially consider the first of them. For the time being, let $j_0 \geq 1$. Then the term $\frac{-u(0)}{h_1}\omega'\left(\frac{r}{h_1}+1\right)$ vanishes. The index j can have two values $j = j_0$ and $j = j_0 - 1$. For the rectangle $Q_{j_0k_0}$, we have $j_0h_1 \leq r \leq (j_0+1)h_1$. Therefore $0 \leq t \leq \frac{r}{h_1} - j \leq 1$, $\omega(t) = t$ and $\omega'(t) = \omega'\left(\frac{r}{h_1} - j\right) = 1$. For $j = j_0 - 1$ we have $\omega'\left(\frac{r}{h_1} - j\right) = -1$. Now the first integral in (1) takes the form

$$\iint\limits_{Q_{j_0k_0}} r\left\{\frac{\partial u}{\partial r} - \frac{1}{h_1}\sum_{k_0-k=0}^{1}\big[u\big((j_0+1)h_1,\ (k+1)h_2\big)-u\big(j_0h_1,\ (k+1)h_2\big)\big]\omega\left(\frac{\vartheta}{h_2}-k\right)\right\}^2 d\vartheta dr\,. \qquad (2)$$

We write $j_0h_1 = r_0$. Using Lagrange's formula (namely, the mean value theorem)

$$u\big((j_0+1)h_1,\ (k+1)h_2\big) - u\big(j_0h_1,\ (k+1)h_2\big) = h_1\frac{\partial}{\partial r}u\big(r_0+\Theta_1h_1,\ (k+1)h_2\big)\,, \quad 0 < \Theta_1 < 1\,,$$

we thereby obtain the integral

$$\iint\limits_{Q_{j_0k_0}} r\left[\frac{\partial}{\partial r}u(r,\vartheta) - \sum_{k_0-k=0}^{1}\frac{\partial}{\partial r}u\big(r_0+\Theta_1h_1,\ (k+1)h_2\big)\omega\left(\frac{\vartheta}{h_2}-k\right)\right]^2 d\vartheta dr\,. \qquad (3$$

We now consider in detail the sum under the integral sign in (3). It has the form

$$\frac{\partial}{\partial r}u\big(r_0+\Theta_1h_1,\ (k_0+1)h_2\big)\omega\left(\frac{\vartheta}{h_2}-k_0\right) + \frac{\partial}{\partial r}u\big(r_0+\Theta_1h_1,\ k_0h_2\big)\omega\left(\frac{\vartheta}{h_2}-k_0+1\right)\,.$$

In the rectangle $Q_{j_0k_0}$, the argument $\frac{\vartheta}{h_2} - k_0$ lies between zero and unity. We set $k_0h_2 = \vartheta_0$ and $\frac{\vartheta}{h_2} = k_0 + \tau$ with $0 \leq \tau \leq 1$. From the definition of ω, we have $\omega\left(\frac{\vartheta}{h_2} - k_0\right) = \tau$, and $\omega\left(\frac{\vartheta}{h_2} - k_0 + 1\right) = 1 - \tau$. By applying Lagrange's formula we transform the expression under the integral sign in (3) to the form

$$\frac{\partial}{\partial r} u(r,\vartheta) - \frac{\partial}{\partial r} u\left(r_0+\Theta_1 h_1, \vartheta_0\right) - \tau h_2 \frac{\partial^2}{\partial r\partial\vartheta} u\left(r_0+\Theta_1 h_1, \vartheta_0+\Theta_2 h_2\right) , \quad 0 < \Theta_2 < 1 . \quad (4)$$

By now applying Lagrange's formula to the first difference in (4), we obtain

$$\left(r-r_0-\Theta_1 h_1\right) \frac{\partial^2}{\partial r^2} u\left(r_0+\Theta_1 h_1+\Theta_3\left(r-r_0-\Theta_1 h_1\right), \vartheta_0\right)$$

$$+ \left(\vartheta-\vartheta_0\right) \frac{\partial^2}{\partial r\partial\vartheta} u\left(r_0+\Theta_1 h_1, \vartheta_0+\Theta_3\left(\vartheta-\vartheta_0\right)\right) - \tau h_2 \frac{\partial^2}{\partial r\partial\vartheta} u\left(r_0+\Theta_1 h_1, \vartheta_0+\Theta_2 h_2\right) , \quad 0 < \Theta_3 < 1 . \quad (5)$$

It can be easily seen that the derivatives $\frac{\partial^2 u}{\partial r^2}$ and $\frac{\partial^2 u}{\partial r\partial\vartheta}$ are bounded in κ . In fact, since $u \in C^{(2)}(\kappa)$, the second derivatives with respect to the cartesian coordinates are bounded. Then the derivatives

$$\frac{\partial^2 u}{\partial r^2} = \frac{\partial^2 u}{\partial x_1^2} \cos^2\vartheta + 2 \frac{\partial^2 u}{\partial x_1 \partial x_2} \cos\vartheta \sin\vartheta + \frac{\partial^2 u}{\partial x_2^2} \sin^2\vartheta \quad (6)$$

and

$$\frac{\partial^2 u}{\partial r\partial\vartheta} = - \frac{\partial u}{\partial x_1} \sin\vartheta + \frac{\partial u}{\partial x_2} \cos\vartheta + r\left[\left(\frac{\partial^2 u}{\partial x_2^2} - \frac{\partial^2 u}{\partial x_1^2}\right) \frac{\sin 2\vartheta}{2} + \frac{\partial^2 u}{\partial x_1 \partial x_2} \cos 2\vartheta\right] \quad (7)$$

are also bounded. Hence, (5) has the estimate $O\left(h_1+h_2\right) \|u\|_{C^{(2)}}$ and the integral (3) has the estimate $h_1 h_2 \|u\|^2_{C^{(2)}} O\left[\left(h_1+h_2\right)^2\right]$.

Now we take the sum of the estimates derived for all the rectangles of the grid for which $j_0 \geq 1$. As a result, we obtain that the first integral in (1), taken over the annulus $h_1 < r < 1$, has the estimate $\|u\|^2_{C^{(2)}} \cdot O\left[\left(h_1+h_2\right)^2\right]$.

It now remains to derive an estimate for the first integral over those rectangles for which $j_0 = 0$. We note that the circular sectors with center at the origin and radius h_1 correspond to the rectangles in the $\left(x_1, x_2\right)$ plane with $j_0 = 0$ and the circle $r < h_1$ corresponds to the totality of these rectangles. The corresponding integrals have the form

$$\iint\limits_{Q_{0,k_0}} r\left[\frac{\partial u}{\partial r} + \frac{u(0)}{h_1} - \frac{1}{h_1} \sum_{\left(-j,k_0-k\right)\in I} u\left(h_1, (k+1)h_2\right) \omega'\left(\frac{r}{h_1} - j_1\right) \omega\left(\frac{\vartheta}{h_2} - k\right)\right]^2 d\vartheta dr . \quad (8)$$

The smallest value of j in (1.6) is zero, and we set $j_0 = 0$. Consequently the condition $(j_0-j,\ k_0-k) \in I$ gives $j = 0$, and the integral (8) takes the simpler form

$$\iint_{Q_{0,k_0}} r\left[\frac{\partial u}{\partial r} + \frac{u(0)}{h_1} - \frac{1}{h_1}\,u\big(h_1,(k_0+1)h_2\big)\omega\left(\frac{\vartheta}{h_2} - k_0\right) - \frac{1}{h_1}\,u\big(h_1,\ k_0h_2\big)\omega\left(\frac{\vartheta}{h_2} - k_0 + 1\right)\right]^2 d\vartheta dr.$$

Here, we have used the facts that $\omega'(t) = 1$, $0 < t < 1$, and k has only two values k_0 and $k_0 - 1$.

On the rectangle Q_{0,k_0}, we have $k_0h_2 < \vartheta < (k_0+1)h_2$; hence $\omega\left(\frac{\vartheta}{h_2} - k_0\right) = \frac{\vartheta}{h_2} - k_0$ and $\omega\left(\frac{\vartheta}{h_2} - k_0 + 1\right) = 1 - \left(\frac{\vartheta}{h_2} - k_0\right)$. The last integral takes the form

$$\iint_{Q_{0,k_0}} r\left\{\frac{\partial u}{\partial r} + \frac{1}{h_1}\left[u(0)-u(h_1,\ k_0h_2)\right] - \frac{1}{h_1}\left[u\big(h_1,\ (k_0+1)h_2\big)-u(h_1,\ k_0h_2)\right]\left(\frac{\vartheta}{h_2} - k_0\right)\right\}^2 d\vartheta dr$$

$$\le 3\iint_{Q_{0,k_0}} r\left(\frac{\partial u}{\partial r}\right)^2 d\vartheta dr + \frac{3}{h_1^2}\iint_{Q_{0,k_0}} r\left[u(0)-u(h_1,\ k_0h_2)\right]^2 d\vartheta dr +$$

$$+ \frac{3}{h_1^2}\iint_{Q_{0,k_0}} r\left[u\big(h_1,\ (k_0+1)h_2\big)-u(h_1,\ k_0h_2)\right]^2 d\vartheta dr = A_1 + A_2 + A_3 . \quad (9)$$

The derivative $\frac{\partial u}{\partial r}$ is bounded and yields $A_1 = O\left(h_1^2h_2\right)\|u\|^2_{C^{(2)}}$. Then, applying Lagrange's formula, we obtain

$$u(0) - u(h_1,\ k_0h_2) = h_1\frac{\partial}{\partial r}u(\Theta h_1,\ k_0h_2) = O(h_1)\|u\|_{C^{(2)}}$$

and hence $A_2 = O\left(h_1^2h_2\right)\|u\|^2_{C^{(2)}}$. Applying Lagrange's formula once more we obtain

$$u\big(h_1,\ (k_0+1)h_2\big) - u(h_1,\ k_0h_2) = h_2\frac{\partial}{\partial\vartheta}u\big(h_1,\ (k_0+\Theta')h_2\big) .$$

From

$$\frac{\partial u}{\partial\vartheta} = -r\left(\frac{\partial u}{\partial x_1}\sin\vartheta - \frac{\partial u}{\partial x_2}\cos\vartheta\right)$$

we have $\frac{\partial u}{\partial\vartheta} = O(r)\|u\|_{C^{(2)}}$. Hence it follows that $A_3 = O\left(h_1^2h_2^3\right)\|u\|^2_{C^{(2)}}$. Now we obtain for (9),

$$A_1 + A_2 + A_3 = O\left(h_1^2h_2+h_1^2h_2^3\right)\|u\|^2_{C^{(2)}} = O\left(h_1^2h_2\right)\|u\|^2_{C^{(2)}} .$$

Taking the sum with respect to k_0, we find

$$\iint\limits_{r<h} r\left(\frac{\partial u}{\partial r} - \frac{\partial u^h}{\partial r}\right)^2 d\vartheta dr = O\left(h_1^2\right)\|u\|^2_{C^{(2)}}$$

which along with the estimate obtained above gives the result

$$\iint\limits_{K} r\left(\frac{\partial u}{\partial r} - \frac{\partial u^h}{\partial r}\right)^2 d\vartheta dr = O\left(\left(h_1+h_2\right)^2\right)\|u\|^2_{C^{(2)}} . \tag{10}$$

§3. Estimation of the Second Integral

We now consider the second integral in (2.1); namely

$$\iint\limits_{Q_{j_0k_0}} \frac{1}{r}\left[\frac{\partial u}{\partial \vartheta} - \frac{1}{h_2}\sum_{(j_0-j,k_0-k)\in I} u\left((j+1)h_1,\ (k+1)h_2\right)\omega\left(\frac{r}{h_1} - j\right)\omega'\left(\frac{\vartheta}{h_2} - k\right)\right]^2 d\vartheta dr . \tag{1}$$

Following the argument of §2, we transform (1) into (where $\vartheta_0 = k_0h_2$)

$$\iint\limits_{Q_{j_0k_0}} \frac{1}{r}\left\{\frac{\partial u}{\partial \vartheta} - \frac{1}{h_2}\sum_{j_0-j=0}^{1} \left[u\left((j+1)h_1,\ \vartheta_0+h_2\right)-u\left((j+1)h_1,\ \vartheta_0\right]\omega\left(\frac{r}{h_1} - j\right)\right\}^2 d\vartheta dr$$

$$= \iint\limits_{Q_{j_0k_0}} \frac{1}{r}\left[\frac{\partial u}{\partial \vartheta} - \sum_{j_0-j=0}^{1} \frac{\partial}{\partial \vartheta} u\left((j+1)h_1,\ \vartheta_0+\Theta h_2\right)\omega\left(\frac{r}{h_1}-j\right)\right]^2 d\vartheta dr . \tag{2}$$

If we put $r = h_1\left(j_0+t\right)$, then $0 \le t \le 1$ on the rectangle Qj_0k_0 . Using the definition of the function ω , we reduce the expression in square brackets in (2) to the form $\left(r_0 = j_0h_1\right)$

$$\frac{\partial}{\partial \vartheta} u(r,\ \vartheta) - \frac{\partial}{\partial \vartheta} u\left(r_0,\ \vartheta_0+\Theta h_2\right) - th_1 \frac{\partial^2}{\partial r\partial \vartheta} u\left(r_0+\Theta_1h_1,\ \vartheta_0+\Theta h_2\right)$$

or, applying Lagrange's formula to the first difference, to

$$\left(r-r_0\right) \frac{\partial^2}{\partial r\partial \vartheta} u\left(r_0+\Theta_2\left(r-r_0\right),\ \vartheta_0+\Theta h_2+\Theta_2\left(\vartheta-\vartheta_0-\Theta h_2\right)\right) + \left(\vartheta-\vartheta_0-\Theta h_2\right) \times$$

$$\times \frac{\partial^2}{\partial \vartheta^2} \left(r_0+\Theta_2\left(r-r_0\right),\ \vartheta_0+\Theta h_2+\Theta_2\left(\vartheta-\vartheta_0-\Theta h_2\right)\right) - th_1 \frac{\partial^2}{\partial r\partial \vartheta} u\left(r_0+\Theta_1h_1,\ \vartheta_0+\Theta h_2\right) . \tag{3}$$

We note that $th_1 = r - r_0$ and that $\dfrac{\partial^2 u}{\partial \vartheta^2}$ is dominated by $Cr\|u\|_{C^{(2)}}$, which follows from

$$\frac{\partial^2 u}{\partial \vartheta^2} = -r\left(\frac{\partial u}{\partial x_1} \cos \vartheta + \frac{\partial u}{\partial x_2} \sin \vartheta\right) + r^2\left(\frac{\partial^2 u}{\partial x_1^2} \sin^2\vartheta - \frac{\partial^2 u}{\partial x_1\partial x_2} \sin^2\vartheta + \frac{\partial^2 u}{\partial x_2^2} \cos^2\vartheta\right) \tag{4}$$

and from the fact that $u \in C^{(2)}(\kappa)$. It is now clear that (2) is dominated by the integral

$$\|u\|^2_{C^{(2)}} \iint\limits_{Q_{j_0,k_0}} \left[\frac{(r-r_0)^2}{r} + r\left(\vartheta-\vartheta_0\right)^2\right] d\vartheta dr \ . \tag{5}$$

If $j_0 \geq 1$, then

$$\iint\limits_{Q_{j_0,k_0}} \frac{(r-r_0)^2}{r} d\vartheta dr = h_2 \int_{r_0}^{r_0+h} \frac{(r-r_0)^2}{r} dr < h_1^2 h_2 \int_{r_0}^{r_0+h} \frac{dr}{r} = h_1^2 h_2 \ln \frac{j_0+1}{j_0}$$

and, analogously,

$$\iint\limits_{Q_{j_0,k_0}} r\left(\vartheta-\vartheta_0\right)^2 d\vartheta dr < h_1 h_2^3 \ .$$

Summing with respect to k_0 for fixed $j_0 \geq 1$ and taking into account that $2n_2 h_2 = 2\pi$, we find that, with $j_0 \geq 1$, the integral over the annulus $j_0 h_1 < r < \left(j_0+1\right)h_1$ is dominated by

$$C\|u\|^2_{C^{(2)}} \left(h_1^2 \ln \frac{j_0+1}{j_0} + h_1 h_2^2\right) .$$

Taking the sum with respect to j_0 from unity to $2n_1 - 2$, we obtain an estimate for the integral over the annulus $h_1 < r < 1$:

$$C\left(h_1^2 \ln\left(2n_1-1\right)+h_2^2\right)\|u\|^2_{C^{(2)}} \leq C\left(h_1^2 \ln \frac{1}{h_1} + h_2^2\right)\|u\|^2_{C^{(2)}} \ . \tag{6}$$

It now remains to derive an estimate for the integral on the circle $r < h_1$. Here, $\omega\left(\frac{r}{h_1}\right) = \frac{r}{h_1}$, $j_0 = 0$, and $j = 0$ or $j = -1$. Since the terms under the integral sign with $j = -1$ vanish, we only have $j = 0$. We fix k_0 , and estimate the integral

$$\iint\limits_{Q_{0,k_0}} \frac{1}{r} \left[\frac{\partial}{\partial\vartheta} u(r, \vartheta) - \frac{r}{h_1 h_2} \sum_{k_0-k=0}^{1} u\left(h_1, (k+1)h_2\right)\omega'\left(\frac{\vartheta}{h_2} - k\right)\right]^2 d\vartheta dr \ . \tag{7}$$

The sum under the integral sign has the form

$$u\left(h_1, \vartheta_0+h_2\right)\omega'\left(\frac{\vartheta}{h_2} - k_0\right) + u\left(h_1, \vartheta_0\right)\omega'\left(\frac{\vartheta}{h_2} - k_0 + 1\right) ,$$

where $\vartheta_0 = k_0 h_2$ and $k_0 h_2 < \vartheta < (k_0+1)h_2$. Then

$$\omega'\left(\frac{\vartheta}{h_2} - k_0\right) = 1 \text{ , for } 0 < \frac{\vartheta}{h_2} - k_0 < 1 \text{ ,}$$

$$\omega'\left(\frac{\vartheta}{h_2} - k_0 + 1\right) = -1 \text{ , for } 1 < \frac{\vartheta}{h_2} - k_0 + 1 < 2 \text{ ,}$$

and we obtain the expression

$$u(h_1, \vartheta_0+h_2) - u(h_1, \vartheta_0) = h_2 \frac{\partial}{\partial\vartheta} u(h_1, \vartheta_0+\Theta h_2) \text{ .}$$

The expression in square brackets under the integral sign (7) takes the form

$$\frac{\partial}{\partial\vartheta} u(r, \vartheta) - \frac{r}{h_1}\frac{\partial}{\partial\vartheta} u(h_1, \vartheta_0+\Theta h_2) \text{ .} \tag{8}$$

Above we obtain that $\frac{\partial u}{\partial\vartheta} = O(r)\|u\|_{C^{(2)}}$. Consequently, (8) is dominated by $Cr\|u\|_{C^{(2)}}$, C = const. But (7) is dominated by

$$C\|u\|^2_{C^{(2)}} \iint\limits_{Q_{0,k_0}} r d\vartheta dr = C\|u\|^2_{C^{(2)}} h_1^2 h_2 \text{ .}$$

Taking the sum with respect to k_0 we see that the integral over the circle $r < h$ is dominated by $O\left(h_1^2\right)\|u\|^2_{C^{(2)}}$. Summing over j_0 and k_0 for the second integral (2.1), we obtain the following estimate $O\left(h_1^2 \ln \frac{1}{h_1} + h_2^2\right)\|u\|^2_{C^{(2)}}$ for it. This along with the formula (2.10) gives the required result

$$\|u-u^h\|_{2,1} = O\left(h_1\sqrt{\ln 1/h_1} + h_2\right)\|u\|_{C^{(2)}} \text{ .} \tag{9}$$

If h_1 and h_2 are infinitesimals of the same order, for example, if $h_1 = ah$, $h_2 = bh$, where a and b are constants, then (9) gives

$$\|u-u^h\|_{2,1} \le Ch\sqrt{\ln 1/h}\ \|u\|_{C^{(2)}} \tag{10}$$

which has slightly worse order than the analogous estimate for rectangular grids (see §§1, 3 of Chapter III). However, as will be shown at the next section, the factor $\sqrt{\ln 1/h}$ can be omitted in many cases.

§4. The Class $C^{(2,\alpha)}$

If $u \in C_0^{(2)}(\kappa) \cap C^{(2,\alpha)}(\kappa)$ and h_1, h_2 are infinitesimals of the same order, then the factor $\sqrt{\ln 1/h}$ in (3.9) can be omitted. It arises only during the estimate of the second integral (2.1). We pause to consider it.

We have $th_1 = r - r_0$ in (3.3), so it can be written as

$$(r-r_0)\left[\frac{\partial^2}{\partial r\partial\vartheta} u\left(r_0+\Theta_2(r-r_0),\ \vartheta_0+\Theta h_2+\Theta_2(\vartheta-\vartheta_0-\Theta h_2)\right) - \frac{\partial^2}{\partial\vartheta\partial r} u\left(r_0+\Theta_1 h_1,\ \vartheta_0+\Theta h_2\right)\right]$$
$$+ \left(\vartheta-\vartheta_0-\Theta h_2\right) \frac{\partial^2}{\partial\vartheta^2} u\left(r_0+\Theta_2(r-r_0),\ \vartheta_0+\Theta h_2+\Theta_2(\vartheta-\vartheta_0-\Theta h_2)\right) . \quad (1)$$

From (2.7) it is clear that $\frac{\partial^2 u}{\partial r\partial\vartheta}$ satisfies a Lipschitz condition with exponent α with respect to r and ϑ, so the first term in (1) is dominated by $O\left[(r-r_0)\cdot\left\{(r-r_0)^\alpha+(\vartheta-\vartheta_0)^\alpha\right\}\right]\|u\|_{C^{(2,\alpha)}}$. From (3.4), the second term in (1) has the estimate $O\left(r(\vartheta-\vartheta_0)\right)\|u\|_{C^{(2,\alpha)}}$. Therefore the integral we are interested in is dominated by

$$\|u\|^2_{C^{(2,\alpha)}} \iint_{Q_{j_0,k_0}} \left[\frac{(r-r_0)^{2+2\alpha}}{r} + \frac{(r-r_0)^2(\vartheta-\vartheta)^{2\alpha}}{r} + r(\vartheta-\vartheta_0)^2\right] d\vartheta dr$$
$$< \left[h_1^2 h_2\left(h_1^{2\alpha}+h_2^{2\alpha}\right)\ln\frac{j_0+1}{j_0} + h_1 h_2^3\right]\|u\|_{C^{(2,\alpha)}} .$$

Taking the sum with respect to k_0, we obtain an estimate for the integral over the annulus $r_0 < r < r_0+h$, it has the form

$$O\left[h_1^2\left(h_1^{2\alpha}+h_2^{2\alpha}\right)\ln\frac{j_0+1}{j_0} + h_1 h_2^2\right]\|u\|^2_{C^{(2,\alpha)}} .$$

Taking the sum with respect to j_0, we derive the estimate

$$O\left[h_1^2\left(h_1^{2\alpha}+h_2^{2\alpha}\right)\ln\frac{1}{h_1} + h_2^2\right]\|u\|^2_{C^{(2,\alpha)}} .$$

The value $\left(h_1^{2\alpha}+h_2^{2\alpha}\right)\ln\frac{1}{h_1}$ is bounded for small values of h_1 and h_2 of the same order, so, finally,

$$\|u-u^h\|_{2,1} \le C\left(h_1+h_2\right)\|u\|_{C^{(2,\alpha)}} . \quad (2)$$

The estimates (3.9) and (2) can be easily transformed to the situation when $u \in C^{(2)}(\kappa)$ or $u \in C^{(2,\alpha)}(\kappa)$, but $u(1, \vartheta) \neq 0$. It is sufficient to extend u to a circle of larger radius so that the extended function belongs to the same class and vanishes on the circumference of the new circle. It is necessary to change the form of the approximating functions.

NOTE. From the estimates of the present chapter, it follows that the coordinate system (1.1) is complete in $W_2^1(\kappa)$.

§5. Approximation in L_p and C

If $u \in C_0^{(2)}(\bar{\kappa})$, it is not difficult to examine the order of the approximation $u^h(x) \approx u(x)$ when $u^h(x)$ is the function (1.5).

Let $r = (j_0+\rho)h_1$, $\vartheta = (k_0+\varphi)h_2$, where j_0, k_0 are integers and $0 \le \rho,\ \varphi \le 1$. Consider the difference

$$u(r, \vartheta) - u^h(r, \vartheta) = u(r, \vartheta) - u(0)\omega\left(\frac{r}{h_1} + 1\right)$$

$$- \sum_{j=0}^{2n_1-2} \sum_{k=-1}^{2n_2-1} u\big((j+1)h_1,\ (k+1)h_2\big)\omega\left(\frac{r}{h_1} - j\right)\omega\left(\frac{\vartheta}{h_2} - k\right) . \quad (1)$$

Initially let $j_0 = 0$, so that $0 \le r \le h$. Then

$$\omega\left(\frac{r}{h_1} + 1\right) = 1 - \frac{r}{h_1} = 1 - \rho , \quad \omega\left(\frac{r}{h_1}\right) = \frac{r}{h_1} = \rho .$$

The terms for which $j \ge 1$ in the sum (1) are equal to zero. Consequently, for a fixed k_0 , the only terms of this sum which are not equal to zero correspond to $k = k_0$ and $k = k_0 - 1$. We have

$$\omega\left(\frac{\vartheta}{h_2} - k_0\right) = \omega(\varphi) = \varphi , \quad \omega\left(\frac{\vartheta}{h_2} - k_0 + 1\right) = \omega(\varphi+1) = 1 - \varphi . \quad (2)$$

Since $j_0 = 0$, the difference (1) takes the form $(\vartheta_0 = k_0h_2)$

$$u(\rho h_1,\ \vartheta_0+\varphi h_2) - u(0)(1-\rho) - u(h_1,\ \vartheta_0+h_2)\rho\varphi - u(h_1,\ \vartheta_0)\rho(1-\varphi) . \quad (3)$$

Taking for (3) a Taylor series expansion about the point $r = 0$, $\vartheta = \vartheta_0$, we find that (3) equals $O\left(h_1^2+h_2^2\right)\|u\|_{C^{(2)}}$ and, hence,

$$|u(r, \vartheta)-u^h(r, \vartheta)| = O\left(h_1^2+h_2^2\right)\|u\|_{C^{(2)}} , \quad r < h_1 . \quad (4)$$

Raising this to the pth power and integrating, we obtain

$$\iint_{r<h_1} |u(r, \vartheta)-u^h(r, \vartheta)|^p r\,dr\,d\vartheta = O\left(h_1^{2p+2}+h_2^{2p}h_1^2\right) . \tag{5}$$

Now let $j_0 \geq 1$. In this situation the formula (1) takes the form

$$u(r, \vartheta) - u^h(r, \vartheta) = u(r, \vartheta) - u\left((j_0+1)h_1, (k_0+1)h_2\right)\left(\frac{r}{h_1} - j_0\right)\left(\frac{\vartheta}{h_2} - k_0\right)$$
$$- u\left(j_0h_1, (k_0+1)h_2\right)\left(j_0 + 1 - \frac{r}{h_1}\right)\left(\frac{\vartheta}{h_2} - k_0\right) - u\left((j_0+1)h_1, k_0h_2\right)\left(\frac{r}{h_1} - j_0\right)\left(k_0 + 1 - \frac{\vartheta}{h_2}\right)$$
$$- u\left(j_0h_1, k_0h_2\right)\left(j_0 + 1 - \frac{r}{h_1}\right)\left(k_0 + 1 - \frac{\vartheta}{h_2}\right) . \tag{6}$$

We put $j_0h_1 = r_0$, $k_0h_2 = \vartheta_0$, so that $r = r_0 + \rho h$, $\vartheta = \vartheta_0 + \varphi h_2$. Taking for the value of u on the r.h.s. of (6) a Taylor series expansion, we obtain for $j_0 \geq 1$,

$$u(r, \vartheta) - u^h(r, \vartheta) = O\left(h_1^2+h_2^2\right)\|u\|_{C^{(2)}} . \tag{7}$$

Raising this to the pth power and integrating over the annulus $h_1 < r < 1$, we derive

$$\iint_{h_1<r<1} |u(r, \vartheta)-u^h(r, \vartheta)|^p r\,dr\,d\vartheta = O\left(h_1^{2p}+h_2^{2p}\right)\|u\|^p_{C^{(2)}} . \tag{8}$$

The sum of (5) and (8) yields the following estimate in L_p :

$$\|u-u^h\|_p = O\left(h_1+h_2\right)\|u\|_{C^{(2)}} . \tag{9}$$

From inequalities (4) and (7), the following estimate in C is obtained

$$\|u-u^h\|_C = O\left(h_1^2+h_2^2\right)\|u\|_{C^{(2)}} . \tag{10}$$

§6. Degenerate Second Order Elliptic Equations

In this final section we review the results of Gusman and Oganesjan [1].

On the rectangle Ω , defined by the inequalities $0 < x < a$, $0 < y < b$, consider the equation

$$- \frac{\partial}{\partial x}\left[p(x, y)\frac{\partial u}{\partial x}\right] - \frac{\partial}{\partial y}\left[y^\alpha q(x, y)\frac{\partial u}{\partial y}\right] = f(x, y) \tag{1}$$

where $0 < \alpha < 1$, $\alpha = \text{const}$, $p, q \in C^{(1)}(\bar{\Omega})$, $p(x, y) \geq \rho_0$, $q(x, y) \geq \rho_0$,

ρ_0 = const. > 0 . We denote the boundary of the rectangle Ω by Γ and the interval $0 \le x \le a$ by Γ' . Consider two problems:

PROBLEM A. *Find the solution of* (1) *in* Ω , *satisfying the condition* $u|_\Gamma = 0$.

PROBLEM B. *Find the solution of* (1) *in* Ω , *satisfying the conditions* $u|_{\Gamma\setminus\Gamma'} = 0$ *and* $\lim_{y\to 0} y^\alpha \frac{\partial u}{\partial y} = 0$.

We note that both problems are positive definite in $L_2(\Omega)$. If $f \in L_2(\Omega)$, then there exist generalised (weak) solutions of both problems. Each of the solutions belong to the corresponding energy space. The scalar product and the norm in both energy spaces are defined by the same formulas

$$[u, \vartheta] = \int_0^a \int_0^b \left[p \frac{\partial u}{\partial x}\frac{\partial \vartheta}{\partial x} + y^\alpha q \frac{\partial u}{\partial y}\frac{\partial \vartheta}{\partial y}\right] dxdy , \tag{2}$$

$$|\!|\!|u|\!|\!|^2 = \int_0^a \int_0^b \left[p\left(\frac{\partial u}{\partial x}\right)^2 + y^\alpha q\left(\frac{\partial u}{\partial y}\right)^2\right] dxdy .$$

The assertions listed above follow from the results of Mikhlin [2] (see also Mikhlin [3]).

We denote by M_A and M_B the sets of functions with the following properties: if $u \in M_A$ or $u \in M_B$, then the functions

$$u,\ u_x,\ u_{x^2},\ u_{x^3},\ y^\alpha u_y,\ \left(y^\alpha u_y\right)_y,\ y^\alpha u_{xy},\ \left(y^\alpha u_y\right)_{xy}$$

are continuous in $\bar{\Omega}$ and satisfy the boundary conditions of problem A or B, respectively.

Assume that problem A (problem B) has a solution $u \in M_A$ $(u \in M_B)$. We construct the approximation u^h as follows. We choose integers n_1, n_2 and put $h_1 = a/n_1$, $h_2 = b/n_2$. We divide the rectangular Ω into $n_1 \cdot n_2$ small rectangles with sides h_1 and h_2 ; we then divide every small rectangle by a diagonal. We take for u^h the function which is linear in each triangle and coincides with the values of u at the vertices. We note that such a construction is equivalent to the construction of $u^h(x, y)$ as a linear combination of coordinate functions generated by the primitive pyramid function (§2, Chapter I), subject to certain linear transforms of the independent variables. For problem A the following inequality holds

$$|\!|\!|u-u^h|\!|\!| \le C\left(h_1^2+h_2^{1-\alpha}\right)^{\frac{1}{2}} \|f\|_{L_2} \tag{3}$$

while for problem B the corresponding inequality is

$$\|u-u^h\| \le C\left(h_1^2+h_2^{2-\alpha}\right)^{\frac{1}{2}}\|f\|_{L_2} \tag{4}$$

both of which guarantee the estimate (3).

CHAPTER VII

APPROXIMATION OF EIGENVALUES[1]

§1. On the Order of the Largest Approximate Eigenvalue. Formulation of the Problem

Let A be a self-adjoint positive definite operator with discrete spectrum, acting in the Hilbert space H, and let $\lambda_1 \le \lambda_2 \le \ldots \le \lambda_n \le \ldots$ be its eigenvalues. We denote by H_A the energy space of the operator A and by $H_A^{(n)}$ any n-dimensional subspace of H_A. Let P_n be the orthogonal projection from H_A onto $H_A^{(n)}$ and let $\lambda_1^{(n)} \le \lambda_2^{(n)} \le \ldots \le \lambda_n^{(n)}$ be the eigenvalues of the operator $P_n A P_n$. It is well known that in the Rayleigh-Ritz method for eigenvalue problems (as well as the variation-difference method), one puts (approximately) $\lambda_\nu = \lambda_\nu^{(n)}$ for $\nu \le n$. Since, for fixed ν, $\lim_{n\to\infty} \lambda_\nu^{(n)} = \lambda_\nu$, it follows that for a fixed ν and sufficiently large n the approximation $\lambda_\nu \approx \lambda_\nu^{(n)}$ is realistic. But, if ν increases along with n, then this approximation may be unrealistic even w.r.t. the order of the growth of $\lambda_\nu^{(n)}$.

We illustrate this with an example. Let $H = L_2(\Omega)$ and let the operator A be defined by

$$Au = -\frac{d}{dx}\, p(x)\, \frac{du}{dx} + q(x)u \ , \quad u(0) = u(1) = 0 \ , \tag{1}$$

where $p \in C^{(1)}[0, 1]$, $p(x) \ge p_0 = \text{const.} > 0$ and the function q is measurable, non-negative and bounded. It is well known, that in this case, λ_n is asymptotically

[1] The results of sections 1-4 are contained in Mikhlin [7-10].

equal to $C_0 n^2$, where C_0 is a positive constant. Hence, it follows that there exist positive constants C', C'' such that $C'n^2 \le \lambda_n \le C''n^2$. We shall formally denote this situation by $\lambda_n \sim n^2$ and use this notation below.

We take as $H_A^{(n)}$ the subspace of polynomials $x(1-x)R_{n-1}(x)$ where $R_{n-1}(x)$ is a polynomial of degree less than of equal to $n - 1$. We shall derive a lower estimate for the largest approximate eignevalue $\lambda_n^{(n)}$. We have

$$\lambda_n^{(n)} = \max_{v \in H_A^{(n)}} \frac{1}{\|v\|^2} \int_0^1 \left[p(x)\left(\frac{dv}{dx}\right)^2 + q(x)v^2\right] dx \ge p_0 \max_{v \in H_A^{(n)}} \frac{\|v'\|^2}{\|v\|^2}, \tag{2}$$

where $\|\cdot\|$ is the norm in the space $L_2(0, 1)$. On the other hand, we have

$$\lambda_n^{(n)} \le p_1 \max \frac{\|v'\|^2}{\|v\|^2} + q_1 \tag{3}$$

where $p_1 = \max p(x)$, $q_1 = \sup q(x)$. From (2) and (3) it follows that

$$\lambda_n^{(n)} \sim \max_{v \in H_A^{(n)}} \frac{\|v'\|^2}{\|v\|^2}. \tag{4}$$

The function $v(x)$ in (4) is a polynomial of degree less than or equal to $n + 1$. By analogy with Markov's L_2-norm inequality for polynomials (see, for example, Bari [1, 2]), we have

$$\max_{v \in H_A^{(n)}} \frac{\|v'\|^2}{\|v\|^2} \sim n^4 .$$

Hence, $\lambda_n^{(n)} \sim n^4$ and $\lambda_n^{(n)} \sim \lambda_n^2$.

It is therefore natural to enquire whether there exist non-trivial[2] subspaces $H_A^{(n)}$ for which the inequality $\lambda_n^{(n)} \le C\lambda_n$ holds, where $C = \text{const}$. An answer to this question is developed in the following sections.

---

[2] We call a subspace *trivial*, if it is spanned by the eigenfunctions of the operator A corresponding to the eigenvalues $\lambda_1, \lambda_2, \ldots, \lambda_n$. For such subspaces, we have $\lambda_\nu^{(n)} = \lambda_\nu$, $\nu = 1, 2, \ldots, n$.

§2. The Rayleigh-Ritz Process

Let A be a positive definite operator with discrete spectrum acting in the Hilbert space H, and let B be an operator semisimilar to A (see Mikhlin [5], §3). We retain the notation of §1. We denote by $\mu_1 \le \mu_2 \le \dots$ the eigenvalues of B, and by φ_n the corresponding orthonormal eigenvunctions in H. We take as $H_A^{(n)}$ the subspace of H_A spanned by $\varphi_1, \varphi_2, \dots, \varphi_n$. We now prove the inequality

$$\lambda_\nu^{(n)} \le C\lambda_\nu \, , \quad \nu = 1, 2, \dots, n \, , \tag{1}$$

where C is independent of n and ν.

Consider the case $\nu = n$. We have

$$\lambda_n^{(n)} = \max_{v \in H_A^{(n)}} \frac{|v|_A^2}{\|v\|^2}$$

where $\|\cdot\|$ denotes the norm in H. Since operators A and B are semi-similar, it follows that

$$|v|_A^2 \sim |v|_B^2 = \sum_{k=1}^{n} \mu_k a_k^2 \, ,$$

where the a_k are the coefficients of the series expansion

$$v = \sum_{k=1}^{n} a_k \varphi_k \, .$$

We have $\|v\|^2 = \sum_{k=1}^{n} a_k^2$, which yields

$$\lambda_n^{(n)} = \max \left(\sum_{k=1}^{n} \mu_k a_k^2 \right) \Big/ \left(\sum_{k=1}^{n} a_k^2 \right) = \mu_n \sim \lambda_n \, .$$

Consequently, there exists a constant C, independent of n, such that $\lambda_n^{(n)} \le C\lambda_n$.

Now let $\nu < n$. For fixed ν and increasing n the value $\lambda_\nu^{(n)}$ does not increase, so $\lambda_\nu^{(n)} \le \lambda_\nu^{(\nu)} \le C\lambda_\nu$. Thus, the inequality (1) follows immediately.

§3. One-Dimensional Variation-Difference Processes

Consider the differential operator, acting in the space $L_2(0, 1)$, and defined by

$$Au = \sum_{k=0}^{s} (-1)^k \frac{d^k}{dx^k}\left(p_k(x)\frac{d^k u}{dx^k}\right) . \tag{1}$$

Let the functions from $\mathcal{D}(A)$ satisfy boundary conditions which guarantee that A is positive definite in $L_2(0, 1)$. We assume also that the energy norm of A is dominated by the norm of the space $W_2^{(s)}(0, 1)$; namely,

$$|\!|\!| u |\!|\!|_A^2 \le C \int_0^1 \sum_{\alpha=0}^{s} [u^{(\alpha)}(x)]^2 dx , \quad C = \text{const.}, \quad \forall u \in H_A , \tag{2}$$

We do not assume that the spectrum of A is *discrete*. The latter does not block the construction of the Rayleigh quotient $|\!|\!| v |\!|\!|_A^2/\|v\|^2$, or the construction of its extremums in corresponding finite-dimensional subspaces. The values of these extremums will be called approximate eigenvalues of the operator A (although this notation can be misleading).

We shall calculate these eigenvalues using variational-difference methods. Let $\omega_q(x)$, $0 \le q \le s-1$, denote a one-dimensional primitive system of degree $s - 1$ with narrow support. We introduce the usual coordinate functions

$$\varphi_{qj}^{(h)}(x) = \omega_q\left(\frac{x}{h} - j\right) , \tag{3}$$

where $h = 1/2n$ (n is an integer) and $-1 \le j \le 2n-1$. The number of functions forming (3) equals $N = s(2n+1)$. The span of these functions is an N-dimensional subspace of H_A, which we denote by $H_A^{(N)}$. We have

$$\lambda_N^{(N)} = \max_{v \in H_A^{(N)}} \frac{|\!|\!| v |\!|\!|_A^2}{\|v\|^2} \tag{4}$$

where $\|\cdot\|$ is the norm in $L_2(0, 1)$. Using the inequality (2) we obtain

$$\lambda_N^{(N)} \le C \max_{v \in H_A^{(N)}} \frac{1}{\|v\|^2} \int_0^1 \sum_{\alpha=0}^{s} [v^{(\alpha)}(x)] dx . \tag{5}$$

We estimate the r.h.s. of (5):

$$\frac{1}{\|v\|^2}\int_0^1 \sum_{\alpha=0}^{s} [v^{(\alpha)}(x)]^2 dx = \left(\sum_{k=0}^{2n-1}\sum_{\alpha=0}^{s}\int_{kh}^{(k+1)h}[v^{(\alpha)}(x)]^2 dx\right)\Bigg/\left(\sum_{k=0}^{2n-1}\int_{kh}^{(k+1)h} v^2(x)dx\right)$$

$$\le \max_{0 \le k \le 2n-1}\left(\sum_{\alpha=0}^{s}\int_{kh}^{(k+1)h}[v^{(\alpha)}(x)]^2 dx\right)\Bigg/\left(\int_{kh}^{(k+1)h} v^2(x)dx\right) ,$$

or, setting $x = (k+t)h$,

$$\frac{1}{\|v\|^2}\int_0^1 \sum_{\alpha=0}^{s} [v^{(\alpha)}(x)]^2 dx \le \max_{0\le k\le 2n-1} \left(\sum_{\alpha=0}^{s} h^{-2\alpha}\int_0^1 [v^{(\alpha)}((k+t)h)]^2 dt\right)\Big/\left(\int_0^1 v^2((k+t)h)dt\right)$$

$$\le h^{-2s} \max_{0\le k\le 2n-1} \left(\sum_{\alpha=0}^{s} \int_0^1 [v^{(\alpha)}((t+k)h)]^2 dt\right)\Big/\left(\int_0^1 v^2((k+t)h)dt\right) . \quad (6)$$

The inclusion $v \in H_A^{(N)}$ implies that

$$v(x) = \sum_{q=0}^{s-1}\sum_{j=-1}^{2n-1} a_{qj}\omega_q\left(\frac{x}{h} - j\right) , \quad a_{qj} = \text{const.}$$

We can therefore rewrite (6) as

$$\frac{1}{\|v\|^2}\int_0^1 \sum_{\alpha=0}^{s} [v^{(\alpha)}(x)]^2 dx \le h^{-2s} \times$$

$$\times \max_{0\le k\le 2n-1} \left(\sum_{\alpha=0}^{s} \int_0^1 \left\{\sum_{q=0}^{s-1} \left[a_{q,k-1}\omega_q^{(\alpha)}(t+1)+a_{qk}\omega_q^{(\alpha)}(t)\right]\right\}^2 dt\right)\Big/$$

$$\Big/\left(\int_0^1 \left\{\sum_{q=0}^{s-1} [a_{q,k-1}\omega_q(t+1)+a_{qk}\omega_q(t)]\right\}^2 dt\right) . \quad (7)$$

The span of the functions $\omega_q(t+1)$, $\omega_q(t)$, $q = 0, 1, \ldots, s-1$, is a finite-dimensional (namely, $2s$-dimensional) subspace of $L_2(0, 1)$. Differentiation operators of order α , $0 \le \alpha \le s$, are bounded on this subspace. Hence, it follows that the quotient on the r.h.s. of (7) is dominated by a constant $\mathcal{D}_s$, and we obtain from (5) and (7) the important estimate

$$\lambda_N^{(N)} \le C\mathcal{D}_s h^{-2s} = O(N^{2s}) . \quad (8)$$

We now assume that the differential operator (1) is non-degenerate: namely, $p_s(x) \ge \tilde{p} = \text{const.} > 0$. In this situation, the spectrum of A , with the usual boundary conditions (essential or natural) is discrete and $\lambda_N \sim N^{2s}$ (for details see Naimark [1]). From this latter result and (8), we obtain the required inequality:

$$\lambda_N^{(N)} \le C\mathcal{D}_s\lambda_N . \quad (9)$$

We denote by d_α the norm of the α-order differentiation operator on the above-mentioned $2s$-dimensional subspace. We can set $\mathcal{D}_s = d_0^2 + d_1^2 + \ldots + d_s^2$. We derive an estimate for d_α when $\omega_q(t)$ are the piecewise polynomial functions of §6, Chapter II. We need an estimate for

$$\int_0^1 [R^{(\alpha)}(t)]^2 dt\Big/\int_0^1 R^2(t)dt , \quad (10)$$

where R is an arbitrary polynomial of degree less than or equal to $2s - 1$. By analogy with Markov's inequality for polynomials in L_2,

$$\int_0^1 [R'(t)]^2 dt \le C_0^2 (2s-1)^4 \int_0^1 R^2(t)dt \ , \quad C_0 = \text{const.}, \tag{11}$$

or equivalently $\|R'\| \le C_0(2s-1)^2\|R\|$. Hence

$$\|R^\alpha\| \le C_0^\alpha[(2s-1)(2s-2)\ \ldots\ (2s-\alpha)]^2\|R\| \tag{12}$$

and it suffices to estimate the constant C_0.

Let $p_r(x)$ be the normalized Legendre polynomial of degree r,

$$p_r(x) = \sqrt{(2r+1)/2}\ P_r(x) \ .$$

We have

$$\int_{-1}^1 [p_r'(x)]^2 dx = \frac{2r+1}{2}\int_{-1}^1 P_r'^2(t)dt$$

$$= \frac{2r+1}{2} P_r(t)P_r'(t)\Big|_{-1}^1 - \frac{2r+1}{2}\int_{-1}^1 P_r(t)P_r''(t)dt = (2r+1)P_r'(1) \ .$$

Using the well known formula $tP_r'(t) - P_{r-1}(t) = rP_r(t)$, we easily find that $P_r'(1) = r(r+1)/2$ and hence,

$$\int_{-1}^1 [p_r'(t)]^2 dt = \frac{r(r+1)(2r+1)}{2} \ .$$

If $\tilde{R}(t)$ is a polynomial of degree less than or equal to $2s - 1$, then we can expand it in terms of normalized Legendre polynomials

$$\tilde{R}(t) = \sum_{r=0}^{2s-1} a_r p_r(t) \ .$$

Then

$$\int_{-1}^1 [\tilde{R}'(t)]^2 dt \le \sum_{r=0}^{2s-1} a_r^2 \sum_{r=0}^{2s-1}\int_{-1}^1 [p_r'(t)]^2 dt$$

$$= \tfrac{1}{2}\int_{-1}^1 \tilde{R}^2(t)dt \sum_{r=0}^{2s-1} r(r+1)(2r+1) = s^2(4s^2-1)\int_{-1}^1 \tilde{R}^2(t)dt \ .$$

Putting $t = 2x - 1$ and $\tilde{R}(2x-1) = R(x)$, we find

$$\int_0^1 [R'(x)]^2 dx \le 4s^2(4s^2-1)\int_0^1 R^2(x)dx \ . \tag{13}$$

From (11) and (13) it follows that we can take C_0^2 as

$$C_0^2 = \frac{4s^2(2s+1)}{(2s-1)^3} . \tag{14}$$

The quotient (14) decreases as s increases. Its maximum value corresponds to $s = 1$ and equals $C_0^2 = 12$. This value is optimal. Indeed, let $s = 1$, so $R(x)$ is a polynomial of degree one: $R(x) = ax + b$. The value C_0^2 must satisfy the inequality $a^2 \le C_0^2\left(\frac{a^2}{3} + ab + b^2\right)$ for any a and b. Letting $a = -2b$, we obtain $C_0^2 \ge 12$.

EXAMPLE. If we apply the variational-difference method to the eigenvalue problem

$$u'' + \lambda u = 0 , \quad u(0) = u(1) = 0 ,$$

and use the primitive function (6.5) of Chapter II, then for the largest approximate eigenvalue (its superscript is $2n-1$) we obtain the estimate $\lambda_{2n-1}^{(2n-1)} \le 12(2n)^2$, which, for large values of n, differs only slightly from the exact eigenvalue

$$\lambda_{2n-1} = \pi^2(2n-1)^2 = 9.8696(2n-1)^2 .$$

We note that the basic results of the present section remain valid when u is a vector function and, hence, the equation (1) is indeed a system of differential equations (see the next section).

§4. The Case of Several Variables

In certain situations, the arguments and results of §3 can be extended with appropriate changes to elliptic operators in multi-dimensional domains. We consider two types of problems:

(1) first boundary value problems for non-degenerate elliptic equations in bounded domains;

(2) an elliptic operator in a parallelapiped.

1. Let A be the following operator acting in $L_2(\Omega)$:

$$\mathrm{A}u = \sum_{|\alpha|=|\beta|=0}^{s} (-1)^{\alpha} D^{\alpha}\left(A_{\alpha\beta}(x) D^{\beta} u\right) , \tag{1}$$

$$u^{(\gamma)}(x) = 0 , \quad |\gamma| \le s-1 , \quad \forall x \in \partial\Omega , \tag{2}$$

where Ω is a bounded domain in $\mathbb{R}_m$. We assume that the Lebesgue measure of the boundary strip Ω_δ tends to zero as its width δ tends to zero. In general, $u(x)$ is a vector-valued function, so the $A_{\alpha\beta}$ are square matrices. Since we assume that

$A_{\alpha\beta} = A^*_{\alpha\beta}$, the differential expression (1) is formally selfadjoint, and the elements of $A_{\alpha\beta}$ are measurable and bounded functions of x in Ω .

We assume that the differential expression (1) is elliptic and non-degenerate in Ω . This implies the existence of a constant $\mu_0 > 0$ such that, for any real numbers t_α and any $x \in \Omega$ the following inequality holds

$$\sum_{|\alpha|=|\beta|=s} A_{\alpha\beta}(x) t_\alpha t_\beta \geq \mu_0 \sum_{|\alpha|=s} t_\alpha^2 . \tag{3}$$

We assume finally that A is positive definite in $L_2(\Omega)$. Using the Friedrichs extension, we can extend A to a selfadjoint operator in $L_2(\Omega)$.

Under the mentioned conditions, the spectrum of A is discrete and positive. Let λ_n denote the eigenvalues of A written in ascending order $0 < \lambda_1 \leq \lambda_2 \leq \dots$. The asymptotic formula for λ_n is well known; namely,

$$\lambda_n \sim n^{2s/m} . \tag{4}$$

For recent papers containing this formula, see, for example, Birman and Solomjak [1]. We shall calculate approximations to λ_n using variational-difference methods. We choose a primitive m-dimensional system of degree $\tilde{s} \geq s-1$ with narrow support. We construct the coordinate functions

$$\varphi_{qj}^{(h)}(x) = \omega_q\left(\frac{x}{h} - j\right) , \quad j \in \hat{J}^h . \tag{5}$$

The number of functions in (5) is finite. Let it be N . Obviously, $N \sim h^{-m}$. We denote by $H_{\mathsf{A}}^{(N)}$, the N-dimensional subspace of the energy space H_{A} spanned by the functions in (5). We have

$$\lambda_N^{(N)} = \max_{v \in H_{\mathsf{A}}^{(N)}} \frac{\|v\|_{\mathsf{A}}^2}{\|v\|^2} = \max_{v \in H_{\mathsf{A}}^{(N)}} \left(\int_{\hat{\Omega}^h} \sum_{|\alpha|=|\beta|=0}^{s} A_{\alpha\beta} v^{(\alpha)} v^{(\beta)} dx \right) \Big/ \left(\int_{\hat{\Omega}^h} v^2 dx \right) . \tag{6}$$

We derive an upper estimate for (6).

Since the matrices $A_{\alpha\beta}(x)$ are bounded, it follows that

$$\lambda_N^{(N)} \leq C \max_{v \in H_{\mathsf{A}}^{(N)}} \left(\int_{\hat{\Omega}^h} \sum_{|\alpha|=0}^{s} [v^{(\alpha)}(x)]^2 dx \right) \Big/ \left(\int_{\hat{\Omega}^h} v^2(x) dx \right) . \tag{7}$$

Using Freidrichs inequality, which can be applied since the conditions of (2) hold, the integrals, containing lower order derivatives, can be estimated by the integrals containing derivatives of order s . This yields

$$\lambda_N^{(N)} \le C \max_{v\in H_A^{(N)}} \left(\int_{\hat{\Omega}^h} \sum_{|\alpha|=s} [v^{(\alpha)}(x)]^2 dx\right) \Big/ \left(\int_{\hat{\Omega}^h} v^2(x)dx\right) . \qquad (8)$$

Let

$$v_0(x) = \sum_{|q|=0}^{\tilde{s}} \sum_{j\in\hat{J}^h} b_{qj}^{(h)} \varphi_{qj}^{(h)}(x) \qquad (9)$$

be the function for which (8) attains its maximum. Then

$$\lambda_N^{(N)} \le C \left(\int_{\hat{\Omega}^h} \sum_{|\alpha|=s} \left[v_0^{(\alpha)}(x)\right]^2 dx\right) \Big/ \left(\int_{\hat{\Omega}^h} v_0^2(x)dx\right) . \qquad (10)$$

We denote by Q_j , the smaller cube of the grid with vertex jh . We have

$$\lambda_N^{(N)} \le C \left(\sum_{j_0\in\hat{J}_0^h} \int_{Q_{j_0}} \sum_{|\alpha|=s} \left[v_0^{(\alpha)}(x)\right]^2 dx\right) \Big/ \left(\sum_{j_0\in\hat{J}_0^h} \int_{Q_{j_0}} v_0^2(x)dx\right)$$

$$\le C \max_{j_0\in\hat{J}^h} \left(\int_{Q_{j_0}} \sum_{|\alpha|=s} \left[v_0^{(\alpha)}(x)\right]^2 dx\right) \Big/ \left(\int_{Q_{j_0}} v_0^2(x)dx\right) . \qquad (11)$$

We put $x = (j_0+t)h$ in the last integral. Then $t \in Q = \{t : \underline{0} \le t \le 1\}$ and the inequality (11) can be easily reduced to

$$\lambda_N^{(N)} \le Ch^{-2s} \max_{j_0\in\hat{J}_0^h} \times$$

$$\left(\int_Q \sum_{|\alpha|=s} \left[\sum_{|q|=0}^{\tilde{s}} \sum_{i\in I} b_{q,j_0-i}\,\omega_q^{(\alpha)}(t+i)\right]^2 dt\right) \Big/ \left(\int_Q \left[\sum_{|q|=0}^{\tilde{s}} \sum_{i\in I} b_{q,j_0-i}\,\omega_q(t+i)\right]^2 dt\right) . \qquad (12)$$

The span of the functions $\omega_q(t+i)$, $0 \le |q| \le s$, $i \in I$, is a finite-dimensional subspace, on which the operator D^α is bounded. Hence, $\lambda_N^{(N)} \le Ch^{-2s} \le CN^{2s/m}$ and, therefore,

$$\lambda_N \le \lambda_N^{(N)} \le C\lambda_N . \qquad (13)$$

2. Now let A be the operator, acting in $L_2(\Omega)$, which is defined by (1), where Q is the cube $\underline{0} \le x \le \underline{1}$. We assume that the differential expression (1) is elliptic, but possibly degenerate in Q . We also assume that:

(a) the boundary conditions, corresponding to (1) are such that A is a positive definite operator in $L_2(\Omega)$;

(b) the energy norm of A is dominated by the norm in $W_2^{(s)}(\Omega)$; namely

$$\|u\|_A \le C\|u\|_{2,s} , \quad \forall u \in H_A . \tag{14}$$

As in subsection 1, we assume that the operator A is extended to be self-adjoint using Friedrichs extension. We do not assume that the spectrum of A is descrete (*cf.* §3).

We choose an integer n and put $h = 1/2n$. We introduce the primitive system and coordinate functions in the same manner as subsection 1, but we replace the set $\hat{\mathcal{J}}^j$ by $\mathcal{J}^h$. In fact, the latter has the form

$$\mathcal{J}^h = \{j : -\underline{1} \le j \le 2\underline{n}-\underline{1}\} . \tag{15}$$

The number of elements of this set and, hence, the number of coordinate functions (5) is equal to (*cf.* (4.1) of Chapter II)

$$N = \binom{\tilde{s}+m-1}{m}(2n+1)^m \sim h^{-m} . \tag{16}$$

As usual, we shall denote the N-dimensional subspace of the space H_A , spanned by the functions in (5), by $H_A^{(N)}$. Calculating the approximate eigenvalues of the operator A in the same way as in §3 we obtain using (14)

$$\lambda_N^{(N)} = \max_{v\in H_A^{(N)}} \frac{\|v\|_A^2}{\|v\|^2} \le C \max_{v\in H_A^{(N)}} \frac{\|v\|_{2,s}^2}{\|v\|^2} = C \max_{v\in H_A^{(N)}} \left(\int_Q \sum_{|\alpha|=0}^{s} [v^{(\alpha)}(x)]^2 dx\right) \Big/ \left(\|v\|^2\right)$$

$$= C\left(\int_Q \sum_{|\alpha|=0}^{s} \left[v_0^{(\alpha)}(x)\right]^2 dx\right) \Big/ \left(\|v_0\|^2\right) . \tag{17}$$

Here, $v_0(x)$ denotes the function from $H_A^{(N)}$ at which the maximum in (17) is attained. This function can be expressed in the form (9) with $\hat{\mathcal{J}}^h$ replaced by $\mathcal{J}^h$ (*cf.* (15)).

Let $\mathcal{J}_0^h$ be the set of labels of low vertices of the smaller grid cubes, lying in Q ; namely, $\mathcal{J}^h = \{j : \underline{0} \le j \le 2\underline{n}-\underline{1}\}$. By (17),

$$\lambda_N^{(N)} \le C\left(\sum_{j_0\in\mathcal{J}_0^h} \int_{Q_{j_0}} \sum_{|\alpha|=0}^{s} \left[v_0^{(\alpha)}(x)\right]^2 dx\right) \Big/ \left(\sum_{j_0\in\mathcal{J}_0^h} \int_{Q_{j_0}} v_0^2(x)dx\right)$$

$$\le C \max_{j_0\in\mathcal{J}_0^h} \left(\int_{Q_{j_0}} \sum_{|\alpha|=0}^{s} \left[v_0^{(\alpha)}(x)\right]^2 dx\right) \Big/ \left(\int_{Q_{j_0}} v_0^2(x)dx\right) .$$

The last inequality can be transformed in the same way as in subsection 1 to

$$\lambda_N^{(N)} \le C \max_{j_0 \in J_0^h}$$
$$\times \left\{\int_Q \sum_{|\alpha|=0}^{s} h^{-2|\alpha|}\left[\sum_{|q|=0}^{\tilde{s}} \sum_{i\in I} b_{q,j_0-i}\omega_q^{(\alpha)}(t+i)\right]^2 dt\right\} \Big/ \left\{\int_Q \left[\sum_{|q|=0}^{\tilde{s}} \sum_{i\in I} b_{q,j_0-i}\omega_q(t+i)\right]^2 dt\right\}$$
$$\le Ch^{-2s} \max_{j_0 \in J_0^h}$$
$$\times \left\{\int_Q \sum_{|\alpha|=0}^{s} \left[\sum_{|q|=0}^{\tilde{s}} \sum_{i\in I} b_{q,j_0-i}\omega_q^{(\alpha)}(t+i)\right]^2 dt\right\} \Big/ \left\{\int_Q \left[\sum_{|q|=0}^{\tilde{s}} \sum_{i\in I} b_{q,j_0-i}\omega_q(t+i)\right]^2 dt\right\}. \quad (18)$$

Since the operators D^α are bounded on the finite-dimensional subspaces, it follows that

$$\lambda_N^{(N)} \le Ch^{-2s} . \quad (19)$$

If the asymptotic estimate $\lambda_N \sim N^{2m/s}$ holds, then

$$\lambda_N^{(N)} \le C\lambda_N . \quad (20)$$

§5. Error Estimate for Fixed Eigenvalues

Such estimates can be obtained using results of Vainikko [1] and from the estimate (3.1) of Chapter III. We present the relevant results of Vainikko. Let T denote a completely-continuous (compact) self-adjoint positive operator, acting in a Hilbert space H. Let the eigenvalues μ_k and eigenfunctions u_k of the operator T be calculated approximately by the energy method with n coordinate functions $\varphi_1, \varphi_2, \ldots, \varphi_n$. Let k be a fixed number, $k < n$, and assume that the eigenvalue μ_k has multiplicity l; namely, $\mu_k = \mu_{k+1} = \ldots = \mu_{k+l-1}$. The approximate eigenvalues will not in general be multiple. We denote them by $\mu_r^{(n)}$, $r = k, k+1, \ldots, k+l-1$. By $u_r^{(n)}$ we denote the approximate eigenfunctions corresponding to $\mu_r^{(n)}$. Let P_k denote the orthogonal projection from H onto the eigen-subspace H_k of T corresponding to the eigenvalue μ_k. Finally, we denote by $E_n(u)$ the best approximation of a function u by linear combinations of the coordinate functions $\varphi_1, \varphi_2, \ldots, \varphi_n$ in H. Then

$$\frac{1}{\mu_r^{(n)}} - \frac{1}{\mu_k} \le C_k \frac{E_n^2\left(P_k u_r^{(n)}\right)}{\left\|P_k u_r^{(n)}\right\|^2} , \quad (1)$$

and

$$\left\|u_r^{(n)}-P_k u_r^{(n)}\right\| \le C_k E_n\left(P_k u_r^{(n)}\right) , \quad C_k = \text{const.} \tag{2}$$

We start with (1). We have $E_n(u) = \|Q^{(n)}u\|$, where $Q^{(n)}$ is the orthogonal projection from H onto the subspace which is orthogonal to $\varphi_1, \varphi_2, \ldots, \varphi_n$. So,

$$\frac{E_n^2\left(P_k u_r^{(n)}\right)}{\left\|P_k u_r^{(n)}\right\|^2} = \left\|Q^{(n)} \frac{P_k u_r^{(n)}}{\left\|P_k u_r^{(n)}\right\|}\right\|^2 .$$

The expression

$$\frac{P_k u_r^{(n)}}{\left\|P_k u_r^{(n)}\right\|}$$

is a normalised eigenfunction of the operator T belonging to the subspace H_k . If $\{u_{k+1}, u_{k+2}, \ldots, u_{k+l-1}\}$ is an orthonormal basis of this subspace, then

$$\frac{P_k u_r^{(n)}}{\left\|P_k u_r^{(n)}\right\|} = \sum_{\rho=k}^{k+l-1} \alpha_\rho u_\rho , \quad \sum_{\rho=k}^{k+l-1} \alpha_\rho^2 = 1 .$$

The quotient on the r.h.s. of (1) takes the form

$$\left\|\sum_{\rho=k}^{k+l-1} \alpha_\rho Q^{(n)} u_\rho\right\|^2$$

which is dominated by

$$\sum_{\rho=k}^{k+l-1} \alpha_\rho^2 \sum_{\rho=k}^{k+l-1} \left\|Q^{(n)} u_\rho\right\|^2 \le l \max_\rho \left\|Q^{(n)} u_\rho\right\|^2 .$$

It now follows from (1) that

$$\frac{1}{\mu_k^{(n)}} - \frac{1}{\mu_k} \le C_k' \max_\rho E_n^2(u_\rho) , \quad C_k' = \text{const.} \tag{3}$$

Let A be a self-adjoint positive-definite operator with discrete spectrum acting in the Hilbert space H . It is known (see Mikhlin [3]) that the application of the energy method to the problem $Au - \lambda u = 0$ in the space H is equivalent to the application of the same method to the problem $\mu u - Tu = 0$, where $\mu = 1/\lambda$, $T = A^{-1}$, in the energy space H_A . Putting $H = H_A$, (3) gives

$$\lambda_n^{(r)} - \lambda_k \le C_k' \max_\rho E_n^2(u_\rho) , \quad r = k, k+1, \ldots, k+l-1 , \tag{4}$$

where this time

$$E_n\left(u_\rho\right) = \left|\mathrm{Q}^{(n)} u_\rho\right|_{\mathrm{A}} \tag{5}$$

and $\mathrm{Q}^{(n)}$ is the orthogonal projection from H_{A} onto the subspace which is orthogonal to $\varphi_1, \varphi_2, \ldots, \varphi_n$.

Let A be a non-degenerate operator of the type described in the preceeding section. We assume that the boundary $\partial\Omega$ and the elements of the matrices $A_{\alpha\beta}$ are sufficiently smooth, then the eigenfunctions of the operator A will be arbitrarily smooth in $\bar{\Omega}$. We assume that the eigenfunctions belong to $W_2^{(2\tilde{s})}(\Omega)$, with $\tilde{s} > s$.

Since A is non-degenerate, the norms in H_{A} and $W_2^{(s)}$ are equivalent; namely, $\left|\mathrm{Q}^{(n)} u_\rho\right|_{\mathrm{A}} \sim \left\|\mathrm{Q}^{(n)} u_\rho\right\|_{2,s}$.

We construct approximate eigenfunctions $u_r^{(n)}$ and eigenvalues $\lambda_r^{(n)}$ by variational-difference methods. For the sake of precision, we assume that the coordinate functions are constructed from the primitive system with narrow support, described in §4 of Chapter I. If the primitive system has wide support or belong to the class described in §8 of Chapter III, then analogous results hold with minor changes. We choose a primitive system of degree $\tilde{s} - 1$. By (3.1) of Chapter III, in which s is replaced by $\tilde{s} - 1$ and $\bar{s}$ by s, we obtain

$$\left\|\mathrm{Q}^{(n)} u_\rho\right\|_{2,s} \le Ch^{\tilde{s}-s}, \tag{6}$$

the required estimate then follows from (4):

$$\lambda_r^{(n)} - \lambda_k = O\left(h^{2(\tilde{s}-s)}\right). \tag{7}$$

Proceeding from (2), it is not difficult to obtain an error estimate for the approximating eigenfunctions. As noted above, with $H = H_{\mathrm{A}}$, (2) can be written as

$$\left|u_r^{(n)} - \mathrm{P}_k^l u_r^{(n)}\right|_{\mathrm{A}} \le C_l E_n\left(\mathrm{P}_k^l u_r^{(n)}\right) = C_k \left|\mathrm{Q}^{(n)} \mathrm{P}_k^l u_r^{(n)}\right|_{\mathrm{A}}. \tag{8}$$

We divide both sides of (8) by $\left|\mathrm{P}_k^l u_r^{(n)}\right|_{\mathrm{A}}$. The expression $u_r^{(n)} / \left|\mathrm{P}_k^l u_r^{(n)}\right|_{\mathrm{A}}$ is an approximate eigenfunction, which we shall again denote by $u_r^{(n)}$. The expression $\mathrm{P}_k^l u_r^{(n)} / \left|\mathrm{P}_k^l u_r^{(n)}\right|$ can be expressed in the form

$$\sum_{\rho=k}^{k+l-1} \alpha_\rho u_\rho .$$

If the eigenfunctions u_ρ are orthonormal in H_A , then

$$\sum_{\rho=k}^{k+l-1} \alpha_\rho^2 = 1 .$$

Using (6), the required estimate easily follows

$$\left\| u_r^{(n)} - u_k \right\|_A = O\left(h^{\tilde{s}-s}\right) . \tag{9}$$

It is not difficult to study the more general eigenvalue problem

$$(A-\lambda B)u = 0 , \tag{10}$$

where A is the operator considered above, and B is a lower order differential operator which is positive definite on the set of functions satisfying conditions (4.2). We do not consider this problem here.

Similar, but slightly less rigorous results, can be found in the book by Varga [1].

CHAPTER VIII

CONSTRUCTION OF VARIATIONAL-DIFFERENCE EQUATIONS

§1. First Boundary Value Problems: Equations with Constant Coefficients on a Cube

Let Q be the cube $\underline{0} \le x \le \underline{1}$ in $\mathbb{R}_m$. Consider the problem

$$\sum_{|\alpha|=|\beta|=0}^{s} (-1)^{\alpha} A_{\alpha\beta} u^{(\alpha+\beta)} = f(x) \ , \tag{1}$$

$$u^{(\gamma)}\big|_{\partial Q} = 0 \ , \quad |\gamma| \le s-1 \ , \tag{2}$$

where the matrices $A_{\alpha\beta} = A^*_{\alpha\beta}$ are constant and the differential operator (1) is elliptic. We assume that the problem (1)-(2) is positive definite in $L_2(Q)$. Let the solution of the problem belong to the space $W_2^{(\tilde{s})}(Q)$, $\tilde{s} > s$, and $f \in C^{(\bar{s})}(Q)$, $\bar{s} \ge 0$. We choose an integer n and put $h = 1/2n$. We then choose a one-dimensional primitive system $\{\omega_r(t)\}$, satisfying the strengthened fundamental completeness conditions (§6, Chapter III). Let s_0 be the smallest integer such that $\tilde{s} \ge 2s_0-1$. Then we take the degree to be $s_0 - 1$. We construct the product primitive functions (see §9, Chapter II):

$$\tilde{\omega}_q(x) = \prod_{k=1}^{m} \omega_{q_k}(x_k) \ , \quad q = (q_1, q_2, \ldots, q_m) \ ,$$

$$0 \le |q| \le \begin{cases} s_0 \ , & s_0 > 1 \ , \\ 0 \ , & s_0 = 0 \ . \end{cases} \tag{3}$$

Finally, we construct the coordinate functions for the variational-difference method as

$$\varphi_{qjh}(x) = \tilde{\omega}_q\left(\frac{x}{h} - j\right) , \quad j \in \hat{\mathcal{J}}^h , \tag{4}$$

with $\hat{\Omega}_h = Q$ and

$$\hat{\mathcal{J}}^h = \{j : \underline{0} \leq j \leq 2\underline{n}-2\} . \tag{5}$$

We seek an approximation to the problem (1)-(2) in the form

$$u_h(x) = \sum_{|q|=0}^{s_0} \sum_{j \in \hat{\mathcal{J}}^h} a_{qj}^{(h)} \varphi_{qjh}(x) . \tag{6}$$

The corresponding algebraic system generated by the variational-difference equations is

$$\sum_{|q'|=0}^{s_0} \sum_{j' \in \hat{\mathcal{J}}^h} [\varphi_{q'j'h}, \varphi_{qjh}] a_{q'j'}^h = (f, \varphi_{qjh}) , \qquad \forall q , \text{ and } \forall j \in \hat{\mathcal{J}}^h . \tag{7}$$

As usual, round brackets denote the scalar product in $L_2(Q)$ and square brackets the scalar product in the energy space of the problem (1)-(2). If $s = 0$, then we set $q = \underline{0}$ in (7).

We examine the matrix and r.h.s. of the system (7) in detail. The energy scalar product becomes

$$[\varphi_{q'j'h}, \varphi_{qjh}] = \int_Q \sum_{|\alpha|=|\beta|=0}^{s} A_{\alpha\beta}\varphi_{qjh}^{(\alpha)}\varphi_{q'j'h}^{(\beta)} dx . \tag{8}$$

Clearly, in order to evaluate the elements of the matrix (7), it is necessary to evaluate the one-dimensional integrals

$$\int_0^1 \omega_r^{(\mu)}\left(\frac{x}{h} - l\right)\omega_{r'}^{(\mu')}\left(\frac{x}{h} - l'\right)dx , \quad l, l' = 0, 1, \ldots, 2n-2 , \qquad \mu, \mu' = 0, 1, \ldots, s , \quad r, r' = 0, 1, \ldots, s_0-1 . \tag{9}$$

The integral (9) fails to vanish only when $l - l' \neq 0, -1, 1$. Since the second and the third situations are identical, it is sufficient to consider $l = l'$ and $l = l' - 1$. In the former, the integrand in (9) fails to vanish only when x lies outside the interval $(lh, (l+2)h)$, so it is sufficient to integrate over this interval. Putting $x = (l+t)h$, we obtain

$$\int_0^2 \omega_t^{(\mu)}(t)\omega_{r'}^{(\mu')}(t)dt \ , \quad \mu, \mu' = 0, 1, \ldots, s \ , \quad r, r' = 0, 1, \ldots, s_0-1 \ . \qquad (10)$$

If $l = l' - 1$, then the integrand in (9) fails to vanish only when x lies outside the interval $\big((l+1)h, (l+2)h\big)$. Putting $x = (l+1+t)h$, we obtain

$$\int_0^1 \omega_r^{(\mu)}(t+1)\omega_{r'}^{(\mu')}(t)dt \ , \quad \mu, \mu' = 0, 1, \ldots, s \ , \quad r, r' = 0, 1, \ldots, s_0-1 \ . \qquad (11)$$

Thus, in order to evaluate the elements of the matrix (7), it is sufficient to evaluate the elementary integrals (10) and (11). In number, they equal $3s_0(s_0+1)(s+1)(s+2)/4$ which is independent of h and m . In addition, the integrals themselves are independent of these parameters. We note that, in some situations, it is only necessary to evaluate a subset of integrals (10) and (11). Thus, for the biharmonic equation, it is only necessary to evaluate those integrals for which the values of μ and μ' are 0 or 2 .

Under the assumption that $f \in C^{(\bar{s})}(Q)$, the right hand side of (7) can be evaluated as follows: we put $x_0 = jh$, $x = x_0 + th$ and expand $f(x)$ into a Taylor series about x_0 :

$$f(x_0+th) = \sum_{|\kappa|=0}^{\bar{s}-1} \frac{h^\kappa}{\kappa!} f^{(\kappa)}(x_0)t^\kappa + O(h^{\bar{s}}) \ .$$

We obtain

$$(f, \varphi_{jqh}) = \int_Q f(x)\tilde{\omega}_q\left(\frac{x}{h} - j\right)dx = \int_{x_0 \underline{<} x \underline{<} x_0+2h} f(x)\omega_q\left(\frac{x}{h} - j\right)dx$$

$$= h^m \int_{0 \underline{<} t \underline{<} 2} f(x_0+th)\tilde{\omega}_q(t)dt \ . \qquad (12)$$

Taking for $f(x_0+th)$ the principal part of the Taylor series, we obtain the following approximate equality, in which the term $O(h^{m+\bar{s}})$ is omitted,

$$(f, \varphi_{qjh}) = \sum_{|\kappa|=0}^{\bar{s}-1} \frac{h^{m+|\kappa|}}{\kappa!} f^{(\kappa)}(x_0) \int_{0 \underline{\le} t \underline{\le} 2} t^\kappa \tilde{\omega}_q(t)dt \qquad (13)$$

where we now have to evaluate new, but fairly elementary, integrals

$$\int_0^2 t^k \omega_r(t)dt \ , \quad r = 0, 1, \ldots, s_0-1 \ , \quad k = 0, 1, \ldots, \bar{s}-1 \ . \qquad (14)$$

We estimate the error associated with the use of (13) to evaluate the right-hand side vector of (7). We denote the vector in question by F^h . Its dimension is $N = O(h^{-m})$. We shall consider F^h as an element of the space $\mathbb{R}_N$.

We denote by δ_{qj}, the difference between the left and right-hand sides of (13) and by δ^h the vector, with components δ_{qj}. As we noted above, $\delta_{qj} = O\left(h^{m+\bar{s}}\right)$, so

$$\|\delta^h\|_{\mathbb{R}_N} = \left[\sum_{|q|=0}^{s_0} \sum_{j\in\hat{J}^h} \delta_{qj}^2\right]^{\frac{1}{2}} = O\left(h^{(m/2)+\bar{s}}\right) . \tag{15}$$

In Chapter IX, we study the influence of this error on the solution of (7) and on the approximate solution of (1)-(2).

§2. First Boundary Value Problems: Equations with Variable Coefficients on a Cube

Consider the following first boundary value problem for the non-degenerate elliptic equations on the cube Q defined above

$$\sum_{|\alpha|=|\beta|=0}^{s} (-1)^{\alpha} D^{\alpha}\left(A_{\alpha\beta} D^{\beta} u\right) = f(x) , \quad A_{\alpha\beta} = A^*_{\beta\alpha} , \quad u^{(\gamma)}|_{\partial\Omega} = 0 , \quad |\gamma| \le s-1 . \tag{1}$$

We assume that the elements of the matrices $A_{\alpha\beta}$ are sufficiently smooth functions of x. We also assume that the operator defined by (1)-(2) is positive definite in $L_2(\Omega)$ and that $f \in C^{(\bar{s})}(Q)$, $\bar{s} \ge 0$. We shall use the primitive system introduced in §1. As before, the calculation of an approximate solution involves the construction and solution of the algebraic system (1.7), the matrix of which is defined by (1.8) and the right-hand side of which is defined by (1.12). The comments of §1 concerning the r.h.s. values and their errors remain valid. Therefore, it is only necessary to consider how the calculation of the matrix elements changes.

Clearly, the term (1.8) remains non-zero only while the components of the vectors $j' - j$ take the values -1, 0 or 1. We put $jh = x_0$, $x = x_0 + th$ and expand the matrices $A_{\alpha\beta}$ using Taylor series to obtain

$$A_{\alpha\beta}\left(x_0+th\right) = \sum_{|\kappa|=0}^{\sigma-1} \frac{h^\kappa}{\kappa!} A_{\alpha\beta}^{(\kappa)}\left(x_0\right) t^\kappa + O\left(h^\sigma\right) . \tag{3}$$

It follows from (3) that an approximate formula for (1.8), with error $O\left(h^{m+\sigma-2s}\right)$, is given by:

$$\begin{aligned}\left[\varphi_{q'j'h}, \varphi_{qjh}\right] &= \sum_{|\kappa|=0}^{\sigma-1} \frac{h^\kappa}{\kappa!} \sum_{|\alpha|=|\beta|=0}^{s} A_{\alpha\beta}^{(\kappa)}\left(x_0\right) \int_Q t^\kappa D^\alpha \tilde{\omega}_{q'}\left(\frac{x}{h} - j'\right) D^\beta \tilde{\omega}_q\left(\frac{x}{h} - j\right) dx \\ &= \sum_{|\kappa|=0}^{\sigma-1} \sum_{|\alpha|=|\beta|=0}^{s} \frac{h^{|\kappa|+m-|\alpha+\beta|}}{\kappa!} A_{\alpha\beta}^{(\kappa)}\left(x_0\right) \int_{0\le t\le 2} t^\kappa \omega_{q'}^{(\alpha)}(t) \omega_q^{(\beta)}(t+j-j') dt .\end{aligned} \tag{4}$$

The evaluation of (1.8) is thereby reduced to the evaluation of one-dimensional integrals of the form

$$\int_0^2 t^k \omega_r^{(\mu)}(t)\omega_{r'}^{(\mu')}(t+\Theta)dt \ , \quad \Theta = -1, 0, 1 \ . \tag{5}$$

It is clear that we should integrate over the interval $(0, 2)$ when $\Theta = 0$; but when $\Theta = 1$ or $\Theta = -1$, the integration need only be taken over $(1, 2)$ or $(0, 1)$, respectively.

We estimate the Eucledian norm of the error associated with approximating the terms in (1.7) by (4). The error is contained in those matrix elements for which $j' - j = -1, 0$ or 1 . The other elements are equal to zero. The number of elements in error is $O(N) = O(h^{-m})$. The error in each element is $O(h^{\sigma+m-2s})$. Denoting the error matrix associated with the use of (4) to evaluate the term in (1.7) by Γ_h , we obtain

$$\|\Gamma_h\|_{\mathbb{R}_N} = O(h^{\sigma+(m/2)-2s}) \ . \tag{6}$$

§3. First Boundary Value Problems: Natural Boundary Conditions

We now examine (2.1) with respect to natural boundary conditions. We shall not formulate them explicitly, but assume that the associated operator is positive definite in $L_2(\Omega)$, that the energy space, considered as a set of elements, coincides with $W_2^{(s)}(\Omega)$, and that the norms in both spaces (energy and $W_2^{(s)}(\Omega)$) are equivalent.

We denote the corresponding solution of (2.1) by $u(x)$. It exists, is unique and belongs to the space $W_2^{(s)}(\Omega)$. If the data associated with (2.1) and the corresponding natural boundary conditions (the boundary conditions, the coefficients and the r.h.s.) are sufficiently smooth, then the solution $u(x)$ will be arbitrary smooth. In fact, we assume that $u \in W_2^{(\tilde{s})}(\Omega)$, $\tilde{s} \geq 2s$.

The function $u(x)$ can be extended to the whole of $\mathbb{R}_m$ so that $u^*(x) \in W_2^{(\tilde{s})}(\mathbb{R}_m)$, and the support of the extended function $u^*(x)$ is compact. Let $\text{supp}\, u^* \subset Q$, where Q is an open cube with sides parallel to the coordinate axes. We construct a grid of grid-size h . Let J_0^h be the set of labels of the lower vertices of the larger cubes contained in Q . Forming products of the one-dimensional primitive functions which yield the highest accuracy of approximation, we construct the product primitive system $\{\tilde{\omega}_q(x)\}$, $|q| \leq s_0$, $s \leq s_0 \leq (\tilde{s}+1)/2$, and the coordinate system

$$\varphi_{qjh}(x) = \tilde{\omega}_q\left(\frac{x}{h} - j\right) .$$

Putting

$$u^h(x) = \sum_{|q|=0}^{s_0} \sum_{j\in\hat{J}_Q^h} h^q u^{*(q)}\big((j+1)h\big)\varphi_{qjh}(x) = \sum_{|q|=0}^{s_0} \sum_{j\in\hat{J}_Q^h} h^q u^{*(q)}\big((j+\underline{1})h\big)\tilde{\omega}_q\left(\frac{x}{h} - j\right) , \tag{1}$$

we obtain from §6, Chapter III, that

$$\|u^*-u^h\|_{W_2^{(s)}(Q)} \le Ch^{2s_0-1-s} \tag{2}$$

where C depends on u but not on h . It follows from (2) that

$$\|u-u^h\|_{W_2^{(s)}(Q)} \le Ch^{2s_0-1-s} \tag{3}$$

and hence, by the equivalence of the energy and $W_2^{(s)}(\Omega)$ norms,

$$|\!|\!|u-u^h|\!|\!| \le Ch^{2s_0-1-s} . \tag{4}$$

For $u^h(x)$ in inequalities (3) and (4), it is sufficient to retain only those terms in its definition (1) for which their supports intersect Ω . Therefore, we can write

$$u^h(x) = \sum_{|q|=0}^{s_0} \sum_{j\in J^h} h^q u^{*(q)}\big((j+\underline{1})h\big)\tilde{\omega}_q\left(\frac{x}{h} - j\right) . \tag{5}$$

We seek an approximate solution of (2.1) with natural boundary conditions in the form

$$u_h(x) = \sum_{|q|=0}^{s_0} \sum_{j\in J^h} a_{qj}^{(h)}\tilde{\omega}_q\left(\frac{x}{h} - j\right) , \quad a_{qj}^{(h)} = \text{const.}, \tag{6}$$

where the coefficients $a_{qj}^{(h)}$ are defined by the requirement that the energy integral of $u - \nu_h$ be minimized, which is equivalent to the requirement that $\|u-u_h\|$ be minimized. It now follows from (4) that

$$|\!|\!|u-u_h|\!|\!| \le Ch^{2s_0-1-s} . \tag{7}$$

Thus, on general domains, the approximate solution of (2.1) with natural boundary condition is simpler than the approximate solution of first boundary value problems. The exception is the cube, for which the first boundary value problem has a simpler approximate solution since (1.6) contains less items than (6) above.

The situation when the boundary conditions are mixed (some are natural while the remainder take the principal form of (2.2)) is also of interest. We limit attention to the simplest case of one-dimensional operators. The results have an obvious extension to multi-dimensional cubes, if a product primitive system is used.

On the interval $0 < x < 1$, consider an ordinary self-adjoint differential equation of order $2s$ (such as (2.1) with $x \in \mathbb{R}_1$) and assume that one of its boundary conditions has the form

$$u^{(k)}(0) = 0 , \quad 0 \le k \le s-1 , \tag{8}$$

and that the remaining boundary conditions are natural. We assume that the operator thereby defined is positive definite in $L_2(0, 1)$. In the one-dimensional case, (5) takes the form

$$u_h(x) = \sum_{q=0}^{s_0} \sum_{j=-1}^{2n-1} h^q u^{(q)}\left((j+1)h\right)\omega_q\left(\frac{x}{h} - j\right) . \tag{9}$$

From (8), the terms with indices $j = -1$, $q = k$ vanish. Thus, it is clear that terms with the mentioned indices can be dropped from $u_h(x)$. In a more general situation, where the boundary conditions include principal conditions of the form

$$u^{(k_1)}(0) = u^{(k_2)}(0) = \ldots = u^{(k_\mu)}(0) = 0 ,$$

$$u^{(l_1)}(1) = u^{(l_2)}(1) = \ldots = u^{(l_\nu)}(1) = 0 ,$$

terms with indices $j = -1$, $q = k_1, k_2, \ldots, k_\mu$ and $j = 2n - 1$, $q = l_1, l_2, \ldots, l_\nu$ vanish. Consequently, the approximate solution can be determined as

$$u_h(x) = \sum_{q=0}^{s_0} \sum_{j=-1}^{2n-1} a_{qj}^{(h)} \omega_q\left(\frac{x}{h} - j\right) ,$$

$$a_{-1,k_1}^{(h)} = a_{-1,k_2}^{(h)} = \ldots = a_{-1,k_\mu}^{(h)} = a_{2n-1,l_1}^{(h)} = a_{2n-1,l_2}^{(h)} = \ldots = a_{2n-1,l_\nu}^{(h)} = 0 . \tag{10}$$

§4. First Boundary Value Problems: Approximation of the Boundary Conditions

As we saw in the preceding sections, it is in general easier to solve first boundary value problems involving natural boundary conditions than either first boundary value problems with or without mixed boundary conditions. As a result, the result, the idea of a "penalty method", in which the principal boundary conditions are approximated by natural ones, would appear to be a possibility. For a bibliography see Oganesjan [1], in which this idea is developed. In the present section, we derive some functional-analytical results related to this problem, without making any claims for the novelty of the results. For simplicity of presentation, we limit attention to the Laplace equation.

Consider the problem

$$-\Delta u = f(x) \ , \quad x \in \Omega \ , \quad f \in L_2(\Omega) \ , \tag{1}$$

$$u\big|_{\partial\Omega} = 0 \ , \tag{2}$$

where Ω is a bounded domain in $\mathbb{R}_m$. For the purposes of the present section it is sufficient to assume that Ω is the union of a finite number of domains each of which is star-shaped with respect to some ball. But, for the sake of the subsequent application of the variational-difference method, we should assume that $\partial\Omega \in C^{(0,1)}$.

We replace the principal boundary condition (2) by the following natural condition

$$\left[\rho \frac{\partial u}{\partial \nu} + u\right]_{\partial\Omega} = 0 \ , \tag{3}$$

where ρ is a sufficiently small positive number. The operator A_ρ defined by (1) and (3) is positive definite in $L_2(\Omega)$. The corresponding energy norm is

$$|\!|\!| u |\!|\!|^2_{A_\rho} = \int_\Omega (\text{grad } u)^2 dx + \frac{1}{\rho}\int_\Gamma u^2 d\Gamma \ , \quad \Gamma = \partial\Omega \ . \tag{4}$$

From (4) it is clear that the operator A_ρ increases as ρ decreases, in the sense that, if $\rho < 1$, which will be assumed below, then $|\!|\!| u |\!|\!|_{A_\rho} \geq |\!|\!| u |\!|\!|_{A_1}$ and we write $A_\rho > A_1$. Since A_1 is positive definite in $L_2(\Omega)$, there exists a constant $\kappa > 0$ such that $|\!|\!| u |\!|\!|_{A_1} \geq \kappa\|u\|$, where $\|\cdot\|$ denotes the norm in $L_2(\Omega)$. Thus, we obtain

$$|\!|\!| u |\!|\!|_{A_\rho} \geq \kappa\|u\| \ , \quad \forall u \in H_{A_\rho} \ . \tag{5}$$

We denote by ϑ_ρ the solution of (1) and (3). From (5), it follows that (see Mikhlin [1], §5)

$$|\!|\!| \vartheta_\rho |\!|\!|_{A_\rho} \leq \frac{1}{\kappa}\|f\| \ , \tag{6}$$

or equivalently

$$\int_\Omega (\text{grad } \vartheta_\rho)^2 dx + \frac{1}{\rho}\int_\Gamma \vartheta_\rho^2 d\Gamma \leq \frac{1}{\kappa^2}\|f\|^2 \ . \tag{6$_1$}$$

Hence,

$$\int_\Gamma \vartheta_\rho^2 d\Gamma \le \frac{\rho}{\kappa^2} \|f\|^2 \tag{7}$$

and therefore, in the norm of $L_2(\Gamma)$, we obtain

$$\lim_{\rho\to\infty} \vartheta_\rho = 0 \; . \tag{8}$$

The inequality (7) can be interpreted as follows: for sufficiently small values of ρ , ϑ_ρ satisfies the boundary condition (2) arbitrarily closely.

From (6), it follows that $\|\vartheta_\rho\|_{A_1} \le \kappa^{-1}\|f\|$. Thus, for $\rho \le 1$, $\|\vartheta_\rho\|_{A_1}$ is bounded independently of ρ . Let ρ be a sequence with a zero limit point. The corresponding set of functions $\vartheta_\rho(x)$ is bounded in H_{A_1} and, hence, in $W_2^{(1)}(\Omega)$. Consequently, this set is weakly compact in $W_2^{(1)}(\Omega)$; that is, there exists a sequence $\{\rho_n\}$ and a function $u_0 \in W_2^{(1)}(\Omega)$ such that ϑ_{ρ_n} converges weakly to u_0 in $W_2^{(1)}(\Omega)$.

We prove that u_0 is the required solution of (1) and (2). Indeed, let η be an arbitrary function from $W_2^{(1)}(\Omega)$. The function ϑ_ρ satisfies the identity

$$\int_\Omega \operatorname{grad} \vartheta_\rho \operatorname{grad} \eta dx + \frac{1}{\rho}\int_\Gamma \vartheta_\rho \eta d\Gamma = (f, \eta) \; . \tag{9}$$

Let $\eta \in C_0^2(\Omega)$, then

$$\int_\Omega \operatorname{grad} \vartheta_\rho \operatorname{grad} \eta dx = \int_\Omega f\eta dx \; .$$

After integration by parts, we obtain

$$-\int_\Omega \vartheta_\rho \Delta\eta dx = \int_\Omega f\eta dx \; .$$

Letting $\rho = \rho_n \to 0$ and using the weak convergence of ϑ_ρ to u_0 in $W_2^{(1)}(\Omega)$, we obtain

$$\int_\Omega u_0 \Delta\eta dx = \int_\Omega f\eta dx \; .$$

This identity implies that, with respect to a given $f \in L_2(\Omega)$, u_0 is the generalized solution of $-\Delta u = f$, so it only remains to prove that $u_0|_\Gamma = 0$. In

(9), let η be an arbitrary function from $C_0^{(2)}(\Omega)'$. We have the weak convergence of ϑ_ρ to u_0 in $W_2^{(1)}(\Omega)$, as $\rho \to 0$, and the strong convergence of ϑ_ρ to "0" in $L_2(\Gamma)$, as $\rho \to 0$. But, the embedding operator mapping the space $W_2^{(1)}(\Omega)$ into the space $L_2(\Gamma)$ is completely continuous. Consequently, the weak convergence of $\vartheta_\rho \to u_0$ in $W_2^{(1)}(\Omega)$ entails the strong convergence of $\vartheta_\rho \to u_0$ in $L_2(\Gamma)$. Hence, it follows that

$$u_0|_\Gamma = 0 \ .$$

The uniqueness of the solution of (1) and (2) along with the weak compactness of the set $\{\vartheta_\rho\}$ in $W_2^{(1)}(\Omega)$, entails the weak convergence of $\vartheta_\rho \to u_0$ in $W_2^{(1)}(\Omega)$. Hence, the required strong convergence of $\vartheta_\rho \to u_0$ in $L_2(\Omega)$ follows.

Stronger results can be obtained under the assumption that the boundary of Ω is sufficiently smooth. More precisely, assume that Γ is defined by $\varphi(x) = 0$ with $\varphi \in C^{(2)}(\bar{\Omega})$ and $\operatorname{grad} \varphi|_\Gamma \neq 0$. In this situation, $u_0 \in W_2^{(2)}(\Omega)$ and using the Sobolev embedding theorems we obtain $\partial u_0/\partial \nu \in L_2(\Gamma)$. In addition

$$\frac{\partial u_0}{\partial \nu} = \sum_{k=1}^{m} \frac{\partial u_0}{\partial x_k} \cos\left(\nu,\ x_k\right) = \pm \frac{1}{\sqrt{|\operatorname{grad} \varphi|^2}} \sum_{k=1}^{m} \frac{\partial u_0}{\partial x_k} \frac{\partial \varphi}{\partial x_k} \ .$$

The right-hand side of this result is defined on a boundary layer Ω_δ (with δ the width of the layer) of Ω and belongs to $W_2^{(1)}\left(\Omega_\delta\right)$. Multiplying the r.h.s. by a continuously differentiable function, which equals unity on the boundary layer $\Omega_{\delta/3}$ and equals zero on $\Omega \backslash \Omega_{2\delta/3}$, we obtain a new function $\psi(x)$ with the following properties: $\psi \in W_2^{(1)}(\Omega)$ and $\psi|_{\partial\Omega} = \partial u_0/\partial \nu$. We put $\vartheta_\rho(x) - u_0(x) = \rho w_\rho(x)$. The function $w_\rho(x)$ is the required solution of

$$\Delta w_\rho = 0 \ , \quad \left[\rho \frac{\partial w_\rho}{\partial \nu} + w_\rho\right]_\Gamma = - \left.\frac{\partial u_0}{\partial \nu}\right|_\Gamma \ .$$

Hence, it yields the minimum of the functional

$$\Phi(w) = \int_\Omega (\operatorname{grad} w)^2 dx + \frac{1}{\rho} \int_\Gamma \left(w + \frac{\partial u_0}{\partial \nu}\right)^2 d\Gamma$$

on $W_2^{(1)}(\Omega)$ (see Mikhlin [3], §2.5). Obviously,

$$\Phi(w_\rho) \le \Phi(-\psi) = \int_\Omega (\text{grad } \psi)^2 dx = c^2 = \text{const.}$$

Therefore

$$\int_\Omega (\text{grad } w_\rho)^2 dx \le c^2 , \tag{10}$$

and

$$\int_\Gamma \left(w_\rho + \frac{\partial u_0}{\partial \nu}\right)^2 d\Gamma \le c^2 \rho . \tag{11}$$

From (11), it follows that $\left\|\vartheta_\rho - u_0 - \rho \frac{\partial u_0}{\partial \nu}\right\|_{L_2(\Gamma)} \le c\rho^{3/2}$ and, hence

$$\|\vartheta_\rho - u_0\|_{L_2(\Gamma)} \le c'\rho . \tag{12}$$

The inequalities (10) and (12) imply that

$$\|\vartheta_\rho - u_0\|_{W_2^{(1)}} = O(\rho) . \tag{13}$$

An error estimate for the "penalty method" is given by (13), when the boundary of the domain is sufficiently smooth.

§5. Variational-Difference Methods on an Axial-Symmetric Grid

We study briefly the construction of variational-difference equation for a plane axial-symmetric grid. We initially consider the following elliptic equation with constant coefficients

$$-\left(A_{11} \frac{\partial^2 u}{\partial x_1^2} + 2A_{12} \frac{\partial^2 u}{\partial x_1 \partial x_2} + A_{22} \frac{\partial^2 u}{\partial x_2^2}\right) + A_0 u = f(x) . \tag{1}$$

In the circle $K = \left\{(x_1, x_2) \ : \ x_1^2 + x_2^2 \le 1\right\}$, we require the solution of the first boundary value problem defined by (1) and the boundary condition

$$u(x)\big|_{\partial K} = 0 . \tag{2}$$

We introduce polar coordinates and construct the axial-symmetric grid with steps h_1 and h_2 as described in §1, Chapter VI. On the basis of (1.5) of Chapter VI, we seek the approximate solution of (1) and (2) in the form

$$u_h(r, \vartheta) = a_0 \omega\left(\frac{r}{h_1} + 1\right) + \sum_{j=0}^{2n_1-2} \sum_{k=-1}^{2n_2-1} a_{jk} \omega\left(\frac{r}{h_1} - j\right) \omega\left(\frac{\vartheta}{h_2} - k\right) , \tag{3}$$

where

$$\omega(t) = \begin{cases} t , & 0 \le t \le 1 , \\ 2-t , & 1 < t \le 2 , \\ 0 , & t \notin [0, 2] . \end{cases}$$

Since we consider (1) and (2) on the unslit circle K , the approximation u_h must be periodic with period 2π with respect to ϑ and, hence, the coefficients in (3) must satisfy

$$a_{j,-1} = a_{j,2n-1} . \tag{4}$$

The scalar product in the energy space of (1) and (2) takes the form

$$[u, v] = \iint\limits_K \left[A_{11} \frac{\partial u}{\partial x_1}\frac{\partial v}{\partial x_1} + A_{12}\left(\frac{\partial u}{\partial x_1}\frac{\partial v}{\partial x_2} + \frac{\partial u}{\partial x_2}\frac{\partial v}{\partial x_1}\right) + A_{22}\frac{\partial u}{\partial x_2}\frac{\partial v}{\partial x_2} + A_0 uv\right]dx_1dx_2$$

or, in polar coordinates, the form

$$\begin{aligned}[u, v] = \int_0^{2\pi}\int_0^1 \Bigg[&\frac{\partial u}{\partial r}\frac{\partial v}{\partial r}\left(A_{11}\cos^2\vartheta + 2A_{12}\cos\vartheta\sin\vartheta + A_{22}\sin^2\vartheta\right) + \\ &+ \frac{1}{r}\left(\frac{\partial u}{\partial r}\frac{\partial v}{\partial \vartheta} + \frac{\partial u}{\partial \vartheta}\frac{\partial v}{\partial r}\right)\left(\left(A_{22}-A_{11}\right)\cos\vartheta\sin\vartheta + A_{12}\left(\cos^2\vartheta-\sin^2\vartheta\right)\right) \\ &+ \frac{1}{r^2}\frac{\partial u}{\partial \vartheta}\frac{\partial v}{\partial \vartheta}\left(A_{11}\sin^2\vartheta - 2A_{12}\cos\vartheta\sin\vartheta + A_{22}\cos^2\vartheta\right) + A_0 uv\Bigg] r\,d\vartheta\,dr .\end{aligned} \tag{5}$$

From (3) and (4), it is clear that an appropriate choice for the coordinate functions is

$$\omega\left(\frac{r}{h_1} + 1\right) , \quad \omega\left(\frac{r}{h_1} - j\right)\omega\left(\frac{\vartheta}{h_2} - k\right) , \quad j = 0, 1, \ldots, 2n_1-2 , \quad k = 0, 1, \ldots, 2n_2-1 .$$

Consequently, the calculation of the terms in the variational-difference matrix reduces to the evaluation with respect to ϑ of integrals over intervals of the form $\left(kh_2, (k+1)h_2\right)$. The functions being integrated are products of 1, ϑ and ϑ^2 with the trigonometric polynomials of degree two occuring in (5). The associated integrals involving r are so simple that they can be evaluated exactly.

With slight modifications, this procedure extends to problems with natural boundary conditions, and to problems defined on circles slit or partially slit along a radius.

From the estimates (3.9) and (4.2) of Chapter VI, we have

$$\|u-u_h\| \le \begin{cases} C\left(h_1\sqrt{\ln 1/h_1} + h_2\right) , & u \in C^{(2)}(\overline{K}) , \\ C\left(h_1+h_2\right) , & u \in C^{(2,\alpha)}(K) , \end{cases} \tag{6}$$

where u denotes the exact solution of (1) and (2), and C a constant.

We now assume that the coefficients of (1) are variable. If it formally remains self-adjoint, then it takes the form

$$- \sum_{j,k=1}^{2} \frac{\partial}{\partial x_j}\left(A_{jk}\frac{\partial u}{\partial x_k}\right) + A_0 u = f(x) \ , \quad A_{21} = A_{12} \ .$$

Naturally, the evaluation of the variational-difference matrix becomes more complicated. Thus, if coefficients of the equation are sufficiently smooth in the unslit closed circle, then they can be approximated using Taylor series expansions in r and trigonometric polynomials in ϑ . Hence, it is only necessary to evaluate integrals of the following functions: products of the functions $1, \vartheta$ and ϑ^2 with trigonometrical polynomials in ϑ , and polynomials in r .

§6. Variational-Difference Schemes Containing a Boundary Layer: The One-Dimensional Situation

A brief discussion of the content of the present section is contained in Mikhlin [18].

Below we only consider primitive functions with a narrow support. From estimates of Chapters II and III we know that: if $\{\omega_q(t)\}$ is an m-dimensional primitive system of degree $s - 1$, satisfying the fundamental completeness condition (or the strengthened fundamental completeness conditions if necessary), and $u \in W^{(\tilde{s})}(\Omega)$, with $\tilde{s} \geq s+1$, then

$$\|u-u^h\|_{\bar{s},\rho} = \|u-u^h\|_{W_\rho^{(\bar{s})}(\Omega)} \leq C\|u\|_{W_\rho^{(\tilde{s})}(\Omega)} h^{\tilde{s}-\bar{s}} \ , \quad 0 \leq \bar{s} \leq s \ ,$$

where

$$u^h(x) = \sum_{|q|=0}^{s-1} \sum_{j\in J^h} A_{qj}^{(h)} \omega_q\left(\frac{x}{h} - j\right)$$

and the coefficients $A_{qj}^{(h)}$ are defined in terms of either the values of the function u and its derivatives or by the corresponding values of the average function. In particular, if $u \in C^{(s-1)}(\bar{\Omega})$ then

$$A_{qj}^{(h)} = h^q u^{(q)}\big((j+1)h\big) \ .$$

Our aim is to show that, for a rather wide range of situations, we can construct another approximating function, which has the same order of approximation, but depending only on the values of the function at grid points and on the values of its derivatives at grid points lying inside some boundary layer. We assume that $u \in C^{(\tilde{s})}(\bar{\Omega})$.

Consider the one-dimensional situation with $\Omega = (0, 1)$. The multidimensional situation will be examined in the next section. If $u \in W_p^{(\tilde{s})}(0, 1)$, $\tilde{s} > s$, then $u \in C^{(s)}[0, 1]$. Setting

$$u^h(x) = \sum_{q=0}^{s-1} \sum_{j=-1}^{2n-1} h^q u^{(q)}\big((j+1)h\big)\omega_q\left(\frac{x}{h} - j\right) \tag{1}$$

we obtain

$$\|u-u^h\|_{p,\bar{s}} \le C_1 h^{\tilde{s}-\bar{s}}, \quad 0 \le \bar{s} \le s, \tag{2}$$

where C_1 depends on u but not on h.

We construct the new approximating functions in the following way: for $u \in C^{(\tilde{s})}$, we select a number σ and coefficients $b_{kl}^{(\sigma)}$ such that the approximate relations

$$h^q u^{(q)}(x) = \sum_{l=-\sigma}^{\sigma} b_{ql}^{(\sigma)} u(x+lh), \quad q = 1, \ldots, s-1, \tag{3}$$

are valid with an accuracy of $O\big(h^{\tilde{s}}\big)$. Using Taylor series, we know that

$$u(x+lh) = \sum_{r=0}^{\tilde{s}-1} l^r h^r \frac{u^{(r)}(x)}{r!} + O\big(h^{\tilde{s}}\big).$$

Consequently, $$\sum_{l=-\sigma}^{\sigma} b_{ql}^{(\sigma)} u(x+lh) = \sum_{r=0}^{\tilde{s}-1} \frac{h^r u^{(r)}\big(x_0\big)}{r!} \sum_{l=-\sigma}^{\sigma} l^r b_{ql}^{(\sigma)} + O\big(h^{\tilde{s}}\big)$$

which yeilds the following system of $\tilde{s}$ equations for the unknown coefficients $b_{ql}^{(\sigma)}$

$$\sum_{l=-\sigma}^{\sigma} l^r b_{ql}^{(\sigma)} = r!\delta_{qr}, \quad r = 0, 1, \ldots, \tilde{s}-1, \tag{4}$$

where $q = 1, 2, \ldots, s-1$. If q is fixed, then the system (4) consists of $\tilde{s}$ equations with $2\sigma + 1$ unknowns. We set $\sigma = (\tilde{s}-1)/2$ if $\tilde{s}$ is odd and $\sigma = \tilde{s}/2$, $b_{q0}^{(\sigma)} = 0$, if $\tilde{s}$ is even. As a result the number of equations equals the number of unknowns and the determinant of the system is a Van der Monde determinant with arguments $-\sigma, -\sigma+1, \ldots, -1, 0, 1, \ldots, \sigma$, for $\tilde{s}$ odd, and arguments $-\sigma, -\sigma+1, \ldots, -1, 1, \ldots, \sigma$, for $\tilde{s}$ even. Thus, the coefficients $b_{ql}^{(\sigma)}$ are uniquely defined, and (3) has sense.

We now transform (1). We have $x \in [0, 1]$, but the function $u(x)$ can be extended to the whole of $\mathbb{R}_1$. Thus, (1) can be rewritten as

$$\hat{u}^h(x) = \sum_{q=0}^{s-1} \sum_{j=-\infty}^{+\infty} h^q u^{(q)}\big((j+1)h\big)\omega_q\left(\frac{x}{h} - j\right) , \quad \forall x \in [0, 1] . \tag{5}$$

In fact, if $x \in [0, 1]$ and $j \le -2$ or $j \ge 2n$, then $\omega_q\left(\frac{x}{h} - j\right)$ vanishes. We set $x = (j+1)h$ in (3) and use it in (5). We then obtain

$$u^h(x) = \check{u}^h(x) + \sum_{q=0}^{s-1} \sum_{j=-\infty}^{\infty} \omega_q\left(\frac{x}{h} - j\right) O\left(h^{\tilde{s}}\right) \tag{6}$$

where

$$\check{u}^h(x) = \sum_{q=0}^{s-1} \sum_{j=-\infty}^{+\infty} \sum_{l=-\sigma}^{\sigma} b_{ql}^{(\sigma)} u\big((j+l+1)h\big)\omega_q\left(\frac{x}{h} - j\right) . \tag{7}$$

For a fixed value of x , at least $2s$ functions $\omega_q\left(\frac{x}{h} - j\right)$ vanish, so the sum in (6) has the estimate $O\left(h^{\tilde{s}}\right)$. The multiplier $O\left(h^{\tilde{s}}\right)$ in this sum does not depend on x . Consequently, the differentiation of (6) k-times with respect to x results in the introduction of the multiplier h^{-k} into (6). Hence we easily see that

$$\|u^h - \check{u}^h\|_{p,\bar{s}} \le C h^{\tilde{s}-\bar{s}} , \quad 0 \le \bar{s} \le s , \tag{8}$$

where C does not depend on h . Comparing this with (3), we obtain

$$\|u - \check{u}^h\|_{p,\bar{s}} \le C h^{\tilde{s}-\bar{s}} , \quad 0 \le \bar{s} \le s . \tag{9}$$

We can simplify (7). We eliminate j by setting $j + l = k$, change the order of summation with respect to k and q and then replace k by j . As a result we obtain

$$\check{u}^h(x) = \sum_{j=-\infty}^{+\infty} u\big((j+1)h\big) \sum_{q=0}^{s-1} \sum_{l=-\sigma}^{\sigma} b_{ql}^{(\sigma)} \omega_q\left(\frac{x}{h} + l - j\right) ,$$

where $b_{00}^{(\sigma)} = 1$ and $b_{0q}^{(\sigma)} = 0$, $q \ne 0$.

We introduce the notation

$$\check{\omega}(t) = \sum_{q=0}^{s-1} \sum_{l=-\sigma}^{\sigma} b_{ql}^{(\sigma)} \omega_q(t+l) . \tag{10}$$

For the new approximations, we obtain the expression

$$\check{u}^h(x) = \sum_{j=-\infty}^{+\infty} u\big((j+1)h\big)\check{\omega}\left(\frac{x}{h} - j\right) . \tag{11}$$

It is clear from (10) that

$$\operatorname{supp} \check{\omega}(x) \subset [-\sigma, \sigma+2] \tag{12}$$

and therefore,

$$\operatorname{supp} \check{\omega}\left(\frac{x}{h} - j\right) \subset [(j-\sigma)h,\ (j+\sigma+2)h] \ . \tag{13}$$

If $j \le -\sigma-2$ and $x \in [0,\ 1]$, then $\frac{x}{h} - j \ge \frac{x}{h} + \sigma + 2 \ge \sigma + 2$ and the function $\check{\omega}\left(\frac{x}{h} - j\right)$ vanishes. Analogously, if $j \ge 2n+\sigma$ and $x \in [0,\ 1]$, then $\check{\omega}\left(\frac{x}{h} - j\right) = 0$. Thus, it is now clear that it suffices to take the sum in (11) for $-\sigma-1 \le j \le 2n+\sigma-1$:

$$\check{u}^h(x) = \sum_{j=-\sigma-1}^{2n+\sigma-1} u\big((j+1)h\big)\check{\omega}\left(\frac{x}{h} - j\right) \ . \tag{14}$$

The expression (14) contains values of u at points lying outside $[0,\ 1]$. The number of these points is independent of h . Thus, if h is sufficiently small, then these points are located in small neighbourhoods of the points $x = 0$ and $x = 1$. The sum of the corresponding terms will be interpreted as a contribution from a boundary layer (this interpretation was used by Sobolev [2] in a similar situation in an examination of cubature formulas). We therefore refer to (14) as an *approximating function with a boundary layer.*

We can construct the boundary layer near only one of the end points; for example, near the point $x = 1$. With this aim in mind, we replace (3) by

$$h^q u^{(q)}(x) \approx \sum_{l=0}^{\tilde{s}-1} C_{ql}^{(\tilde{s})} u(x+lh) \ , \quad q = 1,\ 2,\ \ldots,\ s-1 \ , \tag{15}$$

and define the coefficients $C_{ql}^{(\tilde{s})}$ by

$$C_{0l}^{(\tilde{s})} = \delta_{0l} \ , \quad \frac{1}{r!}\sum_{l=0}^{\tilde{s}-1} C_{ql}^{(\tilde{s})} l^r = \delta_{qr} \ , \quad r = 0,\ 1,\ \ldots,\ \tilde{s}-1 \ , \quad q = 1,\ 2,\ \ldots,\ s-1 \ . \tag{16}$$

Then the error associated with (15) is of $O\left(h^{\tilde{s}}\right)$. Substituting (15) into (5) and neglecting terms of $O\left(h^{\tilde{s}}\right)$, we obtain the new approximation

$$\hat{u}^h(x) = \sum_{j=-\infty}^{+\infty} u\big((j+1)h\big)\hat{\omega}\left(\frac{x}{h} - j\right) \tag{17}$$

where

$$\hat{\omega}(t) = \sum_{q=0}^{s-1}\sum_{l=0}^{\tilde{s}-1} C_{ql}^{(\tilde{s})} \omega_q(t+l) \ . \tag{18}$$

Obviously,

$$\operatorname{supp} \hat{\omega}(t) \subset [-\tilde{s}+1,\ 2] \ . \tag{19}$$

If $j \le -2$ and $x \in [0, 1]$, then $\frac{x}{h} - j \ge 2$ and $\hat{\omega}\left(\frac{x}{h} - j\right) = 0$. Analogously, if $j \ge 2n+l$, then $\frac{x}{h} - j \le \frac{1}{h} - 2n - l = -l$ and $\hat{\omega}\left(\frac{x}{h} - j\right) = 0$. Consequently, it suffices to take the sum in (17) over $-1 \le j \le 2n+l-1$,

$$\hat{u}^h(x) = \sum_{j=-1}^{2n+l-1} u\big((j+1)h\big)\hat{\omega}\left(\frac{x}{h} - l\right) . \tag{20}$$

In (20), the terms with indices $j \ge 2n$ form a boundary layer. The following inequality, analogous to (9), holds for (20):

$$\|u-\hat{u}^h\|_{p,\bar{s}} \le Ch^{\tilde{s}-\bar{s}} , \quad 0 \le \bar{s} \le s , \tag{21}$$

where C is independent of h.

We now turn to the variational-difference method. Consider a boundary value problem for an ordinary differential equation which defines a positive definite operator A in $L_2(0, 1)$. We assume that the corresponding energy norm is dominated by the norm of $W_2^{(s)}(0, 1)$, namely

$$|\!|\!|\vartheta|\!|\!|_{\mathsf{A}} = C\|\vartheta\|_{2,s} , \quad \forall\vartheta \in H_{\mathsf{A}} , \quad C = \text{const.}, \tag{22}$$

where s is an integer. It is then assumed that, for a given f, the solution of $\mathsf{A}u = f$ satisfies $u \in W_2^{(\tilde{s})}(0, 1)$, $\tilde{s} > s$. We take a primitive system $\{\omega_q(t)\}$ of degree $s - 1$ and assume that $u^h(x)$ defined by (1) approximates the solution $u(x)$ and satisfies the inequality (2). We could seek the approximate solution of $\mathsf{A}u = f$ either in the form

$$u_h(x) = \sum_{q=0}^{s-1} \sum_{j=-1}^{2n-1} a_{qj}\omega_q\left(\frac{x}{h} - j\right) \tag{23}$$

or

$$\check{u}_h(x) = \sum_{j=-\sigma-1}^{2n+\sigma-1} \alpha_j\check{\omega}\left(\frac{x}{h} - j\right) . \tag{24}$$

The variational method gives the approximate solution for which the evergy norm of the error is minimized over the examined subspace. Consequently, it follows from (22), (2) and (9) that

$$|\!|\!|u-u_h|\!|\!|_{\mathsf{A}} \le C_1h^{\tilde{s}-s} , \quad |\!|\!|u-\check{u}_h|\!|\!|_{\mathsf{A}} \le C_2h^{\tilde{s}-s} , \tag{25}$$

where C_1 and C_2 are independent of h. Thus, in both cases, the energy norm of the error of the approximate solution has the same order $O\left(h^{\tilde{s}-s}\right)$; but the variational-difference system involves $(2n+1)s$ equations in the first case and only $2(n+\sigma)+1$ in the second. Thus, the solution of the latter is a less laborous matter than the former. It should be noted that the construction of the variational-difference system involves a nearly equal amount of work for both cases.

The application of (24) yields a variational-difference system with band structure. This follows from the fact that supp $\check{\omega}$ is compact. The approximation (24) is analogous to (14). The above results remain valid, if the approximate solution is constructed using the formula which is analogous to (20).

The variational-difference schemes obtained via the use of the approximation (24) and analogous approximations, are called "variational-difference schemes containing a boundary layer".

§7. Variational-Difference Schemes Containing a Boundary Layer: The Multidimensional Situation

For the multidimensional situation, we can construct approximations analogous to (6.14) and (6.20), but containing less parameters than the approximation (1.2) of Chapter II.

Assume that $u \in C^{(\tilde{s})}(\mathbb{R}_m)$ and that the m-dimensional primitive system $\{\omega_q(t)\}$ of degree $s-1$ satisfies the corresponding fundamental completeness conditions so that, if $u^h(x)$ is defined by (1.2) of Chapter II, then

$$\|u-u^h\|_{p,\bar{s}} \le Ch^{s-\bar{s}} \ , \quad 0 \le \bar{s} \le s \ , \quad \|\cdot\|_{p,\bar{s}} = \|\cdot\|_{W_p^{(\bar{s})}(\Omega)} \ , \tag{1}$$

where Ω is a bounded domain in $\mathbb{R}_m$ and C depends on Ω and u, but not on h.

Let $l = (l_1, l_2, \dots, l_m)$ be a multi-index. Consider the problem of determining the constants $b_{ql}^{(\tilde{s})}$, $0 \le |q| \le s-1$, $0 \le |l| \le \tilde{s}-1$, such that

$$h^q u^{(q)}(x) = \sum_{|l|=0}^{\tilde{s}-1} b_{ql}^{(\tilde{s})} u(x+lh) + O(h^{\tilde{s}}) \ , \quad 1 \le |q| \le s-1 \ . \tag{2}$$

Using Taylor series, we obtain

$$u(x+lh) = \sum_{|\beta|=0}^{\tilde{s}-1} \frac{h^\beta l^\beta}{\beta!} u^{(\beta)}(x) + O(h^{\tilde{s}}) \ .$$

Substitution of this in (2) yields

$$h^q u^{(q)}(x) = \sum_{|l|=0}^{\tilde{s}-1} \sum_{|\beta|=0}^{\tilde{s}-1} b_{ql}^{(\tilde{s})} \frac{h^\beta l^\beta}{\beta!} u^{(\beta)}(x) + O\left(h^{\tilde{s}}\right)$$

$$= \sum_{|\beta|=0}^{\tilde{s}-1} \frac{h^\beta}{\beta!} u^{(\beta)}(x) \sum_{|l|=0}^{\tilde{s}-1} b_{ql}^{(\tilde{s})} l^\beta + O\left(h^{\tilde{s}}\right) . \quad (3)$$

The conditions (2) hold, if the coefficients $b_{ql}^{(\tilde{s})}$ are chosen so that

$$\sum_{|l|=0}^{\tilde{s}-1} b_{ql}^{(\tilde{s})} l^\beta = \beta! \delta_{\beta q} , \quad 0 \le |\beta| \le \tilde{s}-1 , \quad 1 \le |q| \le s-1 . \quad (4)$$

For fixed q , (4) is a system of linear equations in which the number of equations equals the number of unknowns and the matrix does not depend on q . We assume that the determinant of the system is non-zero. In some situations, it can be verified directly. Then the coefficients $b_{ql}^{(\tilde{s})}$ are uniquely determined.

Consider (1.2) of Chapter II, and replace the $h^q u^{(q)}\left((j+\underline{1})h\right)$ by the approximations (2). Taking into account the fact that, for any fixed value x , only a finite number of the coordinate functions do not vanish (the number is less than $3^m M_s$, where M_s is the number of multi-indices q for which $|q| \le s-1$), we find

$$u^h(x) = \hat{u}^h(x) + O\left(h^{\tilde{s}}\right) , \quad (5)$$

where the new approximations $\hat{u}^h(x)$ are defined by

$$\hat{u}^h(x) = \sum_{|q|=0}^{s-1} \sum_{j \in J^h} \sum_{|l|=0}^{\tilde{s}-1} b_{ql}^{(\tilde{s})} u\left((j+l+\underline{1})h\right) \omega_q\left(\frac{x}{h} - j\right) . \quad (6)$$

We can prove in the same way as in the preceding section that

$$\|u-\hat{u}^h\|_{p,\bar{s}} \le Ch^{\tilde{s}-\bar{s}} , \quad 0 \le \bar{s} \le s . \quad (7)$$

We can simplify (6) by noting that the summation over $j \in J^h$ can be replaced by the summation over all vectors j with integer components - this can be proved in the same manner as in the preceding section. Consequently, we obtain

$$\hat{u}^h(x) = \sum_{|q|=0}^{s-1} \sum_{j \in J_\infty} \sum_{|l|=0}^{\tilde{s}-1} b_{ql}^{(\tilde{s})} u\left((j+l+\underline{1})h\right) \omega_q\left(\frac{x}{h} - j\right) . \quad (8)$$

After setting $j + l = k$, eliminating j , and then replacing k by j , and introducing the notation

$$\sum_{|q|=0}^{s-1} \sum_{|l|=0}^{\tilde{s}-1} b_{ql}^{(\tilde{s})} \omega_q(t+l) = \hat{\omega}(t) , \quad (9)$$

(3) can be rewritten as

$$\hat{u}^h(x) = \sum_{j\in J_\infty} u\big((j+\underline{1})h\big)\hat{\omega}\left(\frac{x}{h} - j\right) . \tag{10}$$

It is easily proved that

$$\operatorname{supp} \hat{\omega}(t) \subset \{t : -\underline{s}+\underline{1} \le t \le \underline{2}\} . \tag{11}$$

One can also verify that it suffices to take the sum in (10) not over J_∞ but over the set J_+^h , which is the arithmetic sum of the set J^h and the set of multi-indices l , satisfying the inequality $|l| \le \tilde{s}-1$. That is, J_+^h is the set of vectors $j = j' + l$, where $j' \in J^h$ and l is a multi-index such that $|l| \le \tilde{s}-1$. Finally we have

$$\hat{u}^h(x) = \sum_{j\in J_+^h} u\big((j+\underline{1})h\big)\hat{\omega}\left(\frac{x}{h} - j\right) . \tag{12}$$

The terms with indices $j \in J_+^h \setminus J^h$ form the boundary layer in (12).

If problems with natural boundary conditions or defined on a parallelpiped are considered, the reasoning of the preceding section related to variational-difference schemes containing a boundary layer apply without changes to the multidimensional situation. The approximate solution can be sought in the form

$$\hat{u}_h(x) = \sum_{j\in J_+^h} a_j^{(h)}\hat{\omega}\left(\frac{x}{h} - j\right) , \tag{13}$$

where the matrix of the variational-difference system has a band structure.

§8. Non-Self Adjoint Problems

Above, we limited attention to the application of the variational-difference method to self-adjoint positive definite problems. Now we examine its application to a class of non-self-adjoint problems.

Let H be a Hilbert space and A be a self-adjoint positive definite operator in H . Let K be an operator acting in H such that $D(\mathsf{K}) \subset H_{\mathsf{A}}$, where H_{A} is the energy space defined by A , and let the product $\mathsf{T} = \mathsf{A}^{-1}\mathsf{K}$ be completely continuous (compact) in H_{A} . Consider the equation

$$(\mathsf{A}+\mathsf{K})u = f . \tag{1}$$

We take as the solutions of this equation the elements of H_{A} which satisfy (see Mikhlin [3])

$$u + \mathsf{T}u = \mathsf{A}^{-1}f . \tag{2}$$

We assume that, if the solution of (1) is unique, then it exists.

Let $H_A^{(n)}$ be an n-dimensional subspace of the space H_A, $n < \infty$. We apply the Bubnov-Galerkin method to equation (1) (see Mikhlin [3]): we choose a basis $\varphi_1, \varphi_2, \ldots, \varphi_n$ in $H_A^{(n)}$ and put

$$u_n = \sum_{k=1}^{n} a_k\varphi_k \tag{3}$$

where the coefficients a_k are determined by the algebraic system

$$\sum_{k=1}^{n} a_k\{[\varphi_k, \varphi_j]_A + (K\varphi_k, \varphi_j)\} = (f, \varphi_j) \ , \quad j = 1, 2, \ldots, n \, . \tag{4}$$

The system (4) can be easily transformed to

$$\sum_{k=1}^{n} a_k[\varphi_k + T\varphi_k, \varphi_j]_A = \left[A^{-1}f, \varphi_j\right]_A \ , \quad j = 1, 2, \ldots, n \, . \tag{5}$$

It follows from (5) that the application of the Bubnov-Galerkin method to (1) in H with coordinate functions $\varphi_1, \varphi_2, \ldots, \varphi_n$ is equivalent to the application of the same method to (2) in the space H_A with the same coordinate functions.

We denote by P_n the orthogonal-projection operator from H_A into $H_A^{(n)}$. We have $P_n u_n = u_n$, since $u_n \in H_A^{(n)}$. The system (5) is equivalent to

$$u_n + P_n T u_n = P_n A^{-1} f \, . \tag{6}$$

Consider a sequence $\left\{H_A^{(n)}\right\}$ which is complete in the sense of the definition given in §1 of Chapter I. The definition implies that P_n converges strongly to the identity operator in the space H_A

$$|\!|\!| \vartheta - P_n\vartheta |\!|\!|_A \xrightarrow[n\to\infty]{} 0 \ , \quad \forall\vartheta \in H_A \, . \tag{7}$$

Hence, it follows that $\|P_n T - T\|_A \xrightarrow[n\to\infty]{} 0$. Repeating the reasonings of Mikhlin [3], §93, we can prove that (6), and hence the system (5), are uniquely solvable for sufficiently large n and that

$$\lim_{n\to\infty} |\!|\!| u - u_n |\!|\!|_A = 0 \, . \tag{8}$$

An error estimate for $|\!|\!| u - u_n |\!|\!|_A$ follows immediately from Vainikko [1]: $u - u_n = (E + P_n T)^{-1}(u - P_n u)$, where E is the identity operator. For sufficiently large n, the operator $(E + P_n T)^{-1}$ is bounded independently for n. Hence

$$\|u-u_n\|_A \leq C\|u-P_n u\|_A = CE_n(u) \; ; \tag{9}$$

where $E_n(u)$ denotes the best approximation of an element u by elements from $H_A^{(n)}$ with respect to the norm in H_A.

These general considerations will now be applied to the variational-difference method for a differential equation with appropriate boundary conditions. In a bounded domain $\Omega \subset \mathbb{R}_m$, assume that the resulting operator equation can be reduced to the form (1) with A a differential operator of order $2s$ which is self-adjoint and positive definite with respect to the prescribed boundary conditions, and with K another differential operator of order less than s. We assume that the norm in H_A is equivalent to the norm in $W_2^{(s)}(\Omega)$. Then all the conditions imposed above on the operators A and K are satisfied.

We take an m-dimensional primitive system $\{\omega_q(x)\}$ with narrow support and degree $s - 1$, and construct the set $\mathcal{J}^h$ (for a first boundary value problem set $\mathcal{J}^h \equiv \hat{\mathcal{J}}^h$) and put

$$u_h(x) = \sum_{|q|=0}^{s-1} \sum_{j \in \mathcal{J}^h} a_{qj}\varphi_{qjh}(x) \; , \quad \varphi_{qjh}(x) = \omega_q\left(\frac{x}{h} - j\right) . \tag{10}$$

The system (4) becomes

$$\sum_{|q'|=0}^{s-1} \sum_{j' \in \mathcal{J}^h} a_{q'j'}\left\{\left[\varphi_{q'j'h}, \varphi_{qjh}\right] + \left(K\varphi_{q'j'h}, \varphi_{qjh}\right)\right\}$$
$$= \left(f, \varphi_{qjh}\right) \; , \quad |q| \leq s-1 \; , \quad j \in \mathcal{J}^h . \tag{11}$$

If the examined operator equation has a unique solution, then for sufficiently small h the system (11) is uniquely solvable and

$$\lim \|u-u_h\|_A = 0 . \tag{12}$$

If in addition we know that $u \in W_2^{(\tilde{s})}(\Omega)$, $\tilde{s} > s$, then, by (9) and (6.2) of Chapter III, we obtain

$$\|u-u_h\|_A = O\left(h^{\tilde{s}-s}\right) . \tag{13}$$

Above, we examined the application of the simplest, but not necessarily the most appropriate, primitive system. Analogously, we could use a primitive system with wide support, a product primitive system, a system satisfying the strengthened fundamental completeness conditions or a system of higher degree. The formulation of the corresponding results are left to the reader.

CHAPTER IX

ERROR ESTIMATES FOR THE VARIATIONAL-DIFFERENCE METHOD

The present chapter is concerned with the effect of errors, arising during the construction and solution of variational-difference algebraic systems, upon the accuracy of the approximations they generate. A more general formulation for this is studied in Mikhlin [5], where some general theorems are proved and the Ritz process is studied in great detail. It is shown there that the effect of errors, arising during the solution of such systems, is closely connected with the condition number of the corresponding matrices. The latter is well known, especially as a result of the work of Wilkinson [1]. The effect of errors, arising during the construction of such algebraic systems, appears to be closely connected with the notion of stability of the corresponding numerical processes, introduced by Mikhlin.

The stability of variational-difference numerical processes is studied in Mikhlin [7-10]. The corresponding problem has also been considered by Demjanovic [1]. The condition number of variational-difference matrices has been studied by Rukhovetz [1] and Mikhlin [15]. The results of Mikhlin [7-10] are sharpened here.

In §1, we formulate the specified notion of the stability of a numerical process along with relevant theorems (for further details, see Mikhlin [5], Chapters II, VII). In §§2-4, the stability of variational-difference numerical process is studied. In §§5-6, the condition number of variational-difference matrices is examined. A numerical example is given in §7.

§1. On the Stability of Numerical Processes

Let $\{X_n\}$ and $\{Y_n\}$ denote two (not necessaryily countable) sequences of Banach spaces and consider the exact numerical process defined by the following sequence of equations

$$A_n x^{(n)} = y^{(n)} \tag{1}$$

with $x^{(n)} \in X_n$, $y^{(n)} \in Y_n$ and A_n a linear operator mapping X_n onto Y_n. The corresponding non-exact process is defined by the sequence

$$(A_n + \Gamma_n) z^{(n)} = y^{(n)} + \delta^{(n)} . \tag{2}$$

We say that the exact numerical process (1) is *stable* (or more precisely, stable with respect to the sequence of couples $\{(X_n, Y_n)\}$), if there exist positive constants p, q and r, independent of n, such that for any $\delta^{(n)} \in Y_n$ and any Γ_n such that $\|\Gamma_n\|_{X_n \to Y_n} \le r$, the non-exact process (2) is solvable for any n, and the following inequality

$$\|z^{(n)} - x^{(n)}\|_{X_n} \le p\|\Gamma_n\|_{X_n \to Y_n} + q\|\delta^{(n)}\|_{Y_n} \tag{3}$$

holds. If constants p, q and r fail to exist, then the *process* is said to be unstable with respect to the sequence of couples $\{(X_n, Y_n)\}$.

THEOREM 1. *In order that the exact numerical process* (1) *be stable, it is necessary and sufficient that there exist constants* C_1 *and* C_2, *independent of* n, *such that:*

(1) $\left\|A_n^{-1}\right\|_{Y_n \to X_n} \le C$;

(2) *for any sequence of operators* $\{B_n\}$, *with norm equal to unity and mapping* X_n *onto* Y_n, *the following inequality*

$$\left\|A_n^{-1} B_n A_n^{-1} y^{(n)}\right\|_{X_n} \le C_2 .$$

holds.

COROLLARY. *For the stability of the exact numerical process* (1), *it is sufficient that* $\left\|A_n^{-1}\right\|_{Y_n \to X_n}$ *and* $\left\|A_n^{-1} y^{(n)}\right\|_{X_n} = \|x^{(n)}\|_{X_n}$ *be bounded by constants independent of* n.

We now examine the eigenvalue problem. Consider a sequence of Hilbert spaces $\{H_n\}$, and two self-adjoint positive operators A_n and B_n acting in these spaces with the operator A_n positive definite. Let the product $B_n^{\frac{1}{2}}A_n^{-\frac{1}{2}}$ be completely continuous (compact) in H_n. We denote by $\lambda_k^{(n)}$ the kth eigenvalue (in non-decreasing order) of the exact eigenvalue problem

$$(A_n-\lambda B_n)u_n = 0 . \tag{4}$$

As the corresponding non-exact eigenvalue problem, we consider

$$|(A_n+\Gamma_n)-\lambda(B_n+\Delta_n)|\vartheta_n = 0 \tag{5}$$

where Γ_n and Δ_n denote bounded self-adjoint operators acting in H_n. We say that the numerical process for the calculation of the kth eigenvalue of (4) (or more briefly, the process (4)) is *stable* with respect to the sequence $\{H_n\}$, if there exist positive constants p, q and r, independent of n, such that for any Γ_n, with $\|\Gamma_n\| \le r$, and for any Δ_n such that $B_n + \Delta_n > 0$, the following conditions hold:

(1) the operator $A_n + \Gamma_n$ is positive definite in H_n;

(2) the operator

$$(\bar{B}_n+\bar{\Delta}_n)^{\frac{1}{2}}(E_n+\bar{\Gamma}_n)^{\frac{1}{2}}$$

is completely continuous (compact) in H_n, where E_n denotes the identity operator in H_n, $\bar{B}_n = A_n^{-\frac{1}{2}}B_nA_n^{-\frac{1}{2}}$, and the operators $\bar{\Gamma}_n$ and $\bar{\Delta}_n$ are defined in the same way;

(3) if $\mu_k^{(n)}$ is the kth eigenvalue (in non-decreasing order) of (5), then

$$\left|\frac{\lambda_k^{(n)}}{\mu_k^{(n)}} - 1\right| \le p\|\Gamma_n\| + q\|\Delta_n\| . \tag{6}$$

THEOREM 2. *In order that the exact eigenvalue process* (4) *be stable with respect to the sequence of spaces* $\{H_n\}$, *it is necessary that* $\|A_n^{-1}\| \le C_1$ *and sufficient that*

$$\|A_n^{-1}\| \le C_1 , \quad \lambda_k^{(n)} \le C_2 , \quad B_n \le C_3 A_n ,$$

where the constants C_1, C_2, C_3 *are independent of* n.

§2. The Stability of Variational-Difference Processes - The One-Dimensional Problem

Consider the following ordinary differential equation (it is immaterial whether it is degenerate or non-degenerate)

$$\sum_{k=0}^{s} (-1)^k \frac{d^k}{dx^k} \left[p_k(x) \frac{d^k u}{dx^k}\right] = f(x) \ , \quad 0 < x < 1 \ , \tag{1}$$

where the coefficients $p_k(x)$ are assumed to be measurable and bounded. Without specifying the boundary conditions associated with (1), we assume that the operator they determine along with (1) is positive definite in $L_2(0, 1)$, and that the coordinate functions, chosen for the implementation of the variational-difference method, form a complete system in the corresponding energy space.

We apply the variational difference method to the above problem. We consider in detail the case when the primitive system has narrow support and is of degree $s - 1$. Other types of primitive systems, including those associated with variational-difference schemes containing a boundary layer, can be studied in a similar manner and lead to analogous results. The same comments apply to multidimensional problems (see the next section).

The approximations take the form

$$u_h(x) = \sum_{q=0}^{s-1} \sum_{j=-1}^{2n-1} a_{qj}^{(h)} \varphi_{qjh}(x) \ , \quad \varphi_{qjh} = \omega_q\left(\frac{x}{h} - j\right) . \tag{2}$$

When solving the first boundary value problem, it is sufficient to sum over j from 0 to $2n - 2$. This fact does not influence the subsequent reasonings so it will not be used.

The coefficients $a_{qj}^{(h)}$ are determined by the variational-difference algebraic system

$$\sum_{q'=0}^{s-1} \sum_{j'=-1}^{2n-1} a_{q'j'}^{(h)} [\varphi_{q'j'h}, \varphi_{qjh}]_{\mathsf{A}} = (f, \varphi_{qjh}) \ , \quad 0 \le q \le s-1 \ , \quad -1 \le j \le 2n-1 \ , \tag{3}$$

where A is the operator defined by (1) and the associated boundary conditions. For fixed h, the set of coefficients $a_{qj}^{(h)}$, is denoted by the vector $a^{(h)}$ in $\mathbb{R}_N$, $N = 2n - 1$. Similarly, the set of right hand side values of (3) is denoted by the vector $F^{(h)}$, and its components $\left(f, \varphi_{qj}^{(h)}\right)$ by $F_{qj}^{(h)}$.

We derive two auxiliary estimates. Initially, we estimate the norm $\|u_h\| = \|u_h\|_{L_2}$ in terms of $\|a^{(h)}\|_{\mathbb{R}_N}$. We have

$$\|u_h\|^2 = \int_0^1 |u_h(x)|^2 dx = \sum_{l=1}^{2n} \int_{(l-1)h}^{lh} |u_h(x)|^2 dx . \tag{4}$$

Only the functions $\varphi_{q,l-2,h}(x)$ and $\varphi_{q,l-1,h}(x)$ are not identically equal to zero on the interval $((l-1)h, lh)$, and therefore

$$\int_{(l-1)h}^{lh} |u_h(x)|^2 dx = \int_{(l-1)h}^{lh} \left[\sum_{q=0}^{s-1} \left\{a_{q,l-2}^{(h)}\omega_q\left(\frac{x}{h} - l + 2\right) + a_{q,l-1}^{(h)}\omega_q\left(\frac{x}{h} - l + 1\right)\right\}\right]^2 dx$$

$$= h \int_0^1 \left[\sum_{q=0}^{s-1} \left\{a_{q,l-2}^{(h)}\omega_q(t+1)+a_{q,l-1}^{(h)}\omega_q(t)\right\}\right]^2 dt . \tag{5}$$

The last integral is a non-negative form involving the $2s$ variables $a_{q,l-2}^{(h)}$, and $a_{q,l-1}^{(h)}$, $q = 0, 1, \ldots, s-1$. We prove that it is positive definite. Assume that it equals zero. Then

$$\sum_{q=0}^{s-1} \left\{a_{q,l-2}^{(h)}\omega_q(t+1)+a_{q,l-1}^{(h)}\omega_q(t)\right\} \equiv 0 , \quad 0 \le t \le 1 .$$

After differentiating this identity α-times, $\alpha = 0, 1, \ldots, s-1$, and then putting $t = 0$ and $t = 1$, we find that all the coefficients $a_{qj}^{(h)}$ are equal to zero. Hence, it follows that there exist positive constants C_1 and C_2 such that, for any l and h , the following inequalities hold

$$\frac{C_1^2}{2} \sum_{j=l-2}^{l-1} \sum_{q=0}^{s-1} \left[a_{qj}^{(h)}\right]^2 \le \int_0^1 \left[\sum_{q=0}^{s-1} \left\{a_{q,l-2}^{(h)}\omega_q(t+1)+a_{q,l-1}^{(h)}\omega_q(t)\right\}\right]^2 dt$$

$$\le \frac{C_2^2}{2} \sum_{j=l-2}^{l-1} \sum_{q=0}^{s-1} \left[a_{qj}^{(h)}\right]^2 .$$

Taking the sum over all l and using (5), we obtain the required estimate

$$C_1^2 h \sum_{q=0}^{s-1} \sum_{j=-1}^{2n-1} \left[a_{qj}^{(h)}\right]^2 \le \|u_h\|^2 \le C_2^2 h \sum_{q=0}^{s-1} \sum_{j=-1}^{2n-1} \left[a_{qj}^{(h)}\right]^2$$

or more briefly,

$$C_1\sqrt{h}\, \|a^{(h)}\|_{\mathbb{R}_N} \le \|u_h\| \le C_2\sqrt{h}\, \|a^{(h)}\|_{\mathbb{R}_N} . \tag{6}$$

We turn now to the second auxiliary estimate: an estimate of $\|F^{(h)}\|_{\mathbb{R}_N}$ in terms of $\|f\|$. Consider the inner product

$$F_{qj}^{(h)} = (f, \varphi_{qjh}) = \int_0^1 f(x)\varphi_{qjh}(x)dx = \int_{jh}^{(j+2)h} f(x)\omega_q\left(\frac{x}{h} - j\right)dx .$$

Using the Bunyakovsky-Schwartz inequality, we obtain

$$\left[F_{qj}^{(h)}\right]^2 \le \int_{jh}^{(j+2)h} f^2(x)dx \int_{jh}^{(j+2)h} \omega_q^2\left(\frac{x}{h} - j\right)dx = \gamma_q^2 h \int_{jh}^{(j+2)h} f^2(x)dx ,$$

$$\gamma_q^2 = \int_0^2 \omega_q^2(t)dt .$$

Taking the sum over q and j, we derive the second auxiliary estimate

$$\|F^{(h)}\|_{\mathbb{R}_N} \le \gamma\sqrt{h}\ \|f\| , \quad \gamma^2 = 2\sum_{q=0}^{s-1} \gamma_q^2 . \tag{7}$$

We now introduce the N-dimensional spaces X_N and Y_N : if b is a vector in such a space with components b_{qj} , $q = 0, 1, \ldots, s-1$, $j = -1, 0, \ldots, 2n-1$, then we set

$$\|b\|_{X_N} = \sqrt{h}\ \|b\|_{\mathbb{R}_N} , \quad \|b\|_{Y_n} = \frac{1}{\sqrt{h}} \|b\|_{\mathbb{R}_N} . \tag{8}$$

The spaces X_N and Y_N are Hilbert spaces which are conjugate with respect to the inner product in $\mathbb{R}_n$. The inequalities (6) and (7) can be rewritten using the new norms as

$$C_1\|a^{(h)}\|_{X_N} \le \|u_h\| \le C_2\|a^{(h)}\|_{X_N} \tag{6$_1$}$$

and

$$\|F^h\|_{Y_N} \le \|f\| . \tag{7 }$$

We denote by M_h the matrix of (4), and by A_h the corresponding operator mapping X_N onto Y_N . The vectors $a^{(h)}$ and $F^{(h)}$ can be examined as elements of the spaces X_n and Y_n , respectively. Then the vector $a^{(h)}$ is determined as the solution of the equation

$$A_h a^{(h)} = F^{(h)} . \tag{9}$$

THEOREM 1. *The numerical process* (9) (*the variational-difference process for the determination of the coefficients* $a_{qj}^{(h)}$) *is stable with respect to the sequence of spaces* (X_N, Y_N) .

We derive an estimate for the norm $\left\|A_h^{-1}\right\|$. We have

$$\left\|A_h^{-1}\right\| = \sup_{b\in \mathbb{R}_N} \frac{\left\|A_h^{-1}b\right\|_{X_N}}{\|b\|_{Y_N}} = h \sup_{b\in \mathbb{R}_N} \frac{\left\|M_h^{-1}b\right\|_{\mathbb{R}_N}}{\|b\|_{\mathbb{R}_N}} = \frac{h}{\mu_1^{(N)}} \tag{10}$$

where $\mu_1^{(N)}$ is the smallest eigenvalue of the matrix M_h.

Writing

$$\lambda_0 = \inf_{\vartheta\in H_A} \frac{|\!|\!|\vartheta|\!|\!|_A^2}{\|\vartheta\|^2}$$

we have that $\lambda_0 > 0$, since A is positive definite. We denote by $\vartheta^h(x)$ an arbitrary function of the form

$$\vartheta^h(x) = \sum_{q=0}^{s-1} \sum_{j=-1}^{2n-1} b_{qj}\varphi_{qjh}(x)$$

and by b a vector in $\mathbb{R}_N$ with components b_{qj}. Obviously, since $\vartheta^h \in H_A$,

$$\lambda_0 \le \frac{\|\vartheta^h\|_A^2}{\|\vartheta^h\|^2} .$$

Further,

$$|\!|\!|\vartheta^h|\!|\!|_A^2 = \left|\!\left|\!\left| \sum_{q=0}^{s-1} \sum_{j=-1}^{2n-1} b_{qj}\varphi_{qjh} \right|\!\right|\!\right|_A^2 = \sum_{q,q'=0}^{s-1} \sum_{j,j'=-1}^{2n-1} b_{qj} b_{q'j'} [\varphi_{qjh}, \varphi_{q'j'h}]_A = (M_h b, b)_{\mathbb{R}_N} .$$

Besides, by (6), we have $\|\vartheta^h\|^2 \ge C_1^2 h \|b\|_{\mathbb{R}_N}^2$. Hence

$$\lambda_0 \le \frac{1}{C_1^2 h} \frac{(M_h b, b)_{\mathbb{R}_N}}{\|b\|_{\mathbb{R}_N}^2} .$$

Taking the infimum of the right hand side with respect to all possible vectors $b \in \mathbb{R}_N$, we find that $\lambda_0 \le \mu_1^{(N)}/C_1^2 h$; that is, $\mu_1^{(N)} \ge \lambda_0 C_1^2 h$. Using this in (10), we obtain $\left\|A_h^{-1}\right\| \le 1/\lambda_0 C_1^2$, which corresponds to condition (1) of Theorem 1 of §1.

The variational-difference process for the above problem converges in H_A. This is an automatic consequence of the completeness of the coordinate system. Therefore, this process converges in $L_2(0, 1)$. Hence, the L_2-norms of the approximate solutions are bounded: $\|u_h\| \le C_3 = \text{const}$. By the inequality ($6_1$), the norms $\|a^{(n)}\|_{X_N}$ are also bounded: $\|a^{(h)}\|_{X_N} \le C_3/C_1$. The corollary of §1 implies the stability of the variational-difference process for the determination of the required coefficients in the sequence of spaces (X_N, Y_N). It yields the following interpretation.

Let the matrix M_h be evaluated with error Γ_h and the r.h.s.-vector $F^{(h)}$ with the error $\delta^{(h)}$. Let $c^{(h)}$ be the solution of the nonexact system

$$\left(A_h+\tilde{\Gamma}_h\right)c^{(h)} = F^{(h)} + \delta^{(h)} , \tag{11}$$

where $\tilde{\Gamma}_h$ is the operator mapping X_N onto Y_N induced by the matrix Γ_h. Then there exist positive constants p, q and r such that, for $\|\tilde{\Gamma}_h\| \le r$, the following inequality holds

$$\|c^{(h)}-a^{(h)}\| \le p\|\tilde{\Gamma}_h\| + q\|\delta^{(h)}\|_{Y_N} . \tag{12}$$

We have studied the stability of the numerical process for the determination of the coefficients in (2). This formula defines the approximate solution of the above problem. It is not difficult to describe the numerical process which determines the approximate solution itself. Formula (2) defines the operator mapping X_N onto $H_A^{(N)}$ and bringing the vector of coefficients a^h into correspondence with the approximate solution $u_h(x)$. We denote this operator by Π_h. Obviously, it can be inverted; that is, $u_h = \Pi_h a^{(h)}$ becomes $a^{(h)} = \Pi_h^{-1} u_h$. Substituting this into (9), we obtain the numerical process for u_h itself:

$$A_h \Pi_h^{-1} u_h = F^{(h)} . \tag{13}$$

THEOREM 2. *The numerical process* (13) *is stable with respect to the sequence of spaces* $\left(H_A^{(N)}, Y_N\right)$.

We recall that $H_A^{(N)}$ is the subspace of H_A spanned by the functions in (2).

Along with the "exact" equation (13), we consider the "nonexact" equation

$$\left(A_h+\tilde{\Gamma}_h\right)\Pi_h^{-1}\vartheta_h = F^{(h)} + \delta^{(h)} , \tag{14}$$

where the "nonexact approximate solution" is given by

$$\vartheta_h(x) = \sum_{q=0}^{s-1} \sum_{j=-1}^{2n-1} c_{qj}^{(h)} \varphi_{qjh}(x) , \tag{15}$$

where the $c_{qj}^{(h)}$ denote the components of the vector $c^{(h)}$ satisfying (11). We have

$$\| \vartheta_h - u_h \|_A^2 = \left(M_h \left(c^{(h)} - a^{(h)}\right), \; c^{(h)} - a^{(h)} \right)_{\mathbb{R}_N} \le \left\| A_h \left(c^{(h)} - a^{(h)}\right) \right\|_{Y_N} \cdot \| c^{(h)} - a^{(h)} \|_{X_N} . \tag{16}$$

The second multiplier is estimated by (12). We derive an estimate for the first. Together equation (9) and the formula (11) imply

$$\left(A_h + \tilde{\Gamma}_h\right) \left(c^{(h)} - a^{(h)}\right) = \delta^{(h)} - \tilde{\Gamma}_h a^{(h)} .$$

The last equality can be reduced to the form

$$\left(E + \tilde{\Gamma}_h A_h^{-1}\right) A_h \left(c^{(h)} - a^{(h)}\right) = \delta^{(h)} - \tilde{\Gamma}_h a^{(h)} ,$$

where E denotes the identity operator. As we saw above, $\left\| A_h^{-1} \right\| \le 1/C_1^2 \lambda_0$. We stipulate that $\| \tilde{\Gamma}_h \| \le \beta C_1^2 \lambda_0$, where β is any number from the interval $(0, 1)$.

Then $\left\| \left(E + \tilde{\Gamma}_h A_h^{-1}\right)^{-1} \right\| \le (1-\beta)^{-1}$ and

$$\left\| A_h \left(c^{(h)} - a^{(h)}\right) \right\|_{Y_N} \le \frac{1}{1-\beta} \| \delta^h \|_{Y_N} + \frac{C_3}{(1-\beta) C_1} \| \tilde{\Gamma}_h \| .$$

Now, from (16) and (12), we obtain

$$\| \vartheta_h - u_h \|_A \le \left\{ \frac{1}{1-\beta} \left[p \| \tilde{\Gamma}_h \| + q \| \delta^{(h)} \|_{Y_N} \right] \left[\frac{C_3}{C_1} \| \tilde{\Gamma}_h \| + \| \delta^{(h)} \|_{Y_N} \right] \right\}^{\frac{1}{2}} \le p_1 \| \tilde{\Gamma}_h \| + q_1 \| \delta^h \|_{Y_N} , \tag{17}$$

provided that $\| \tilde{\Gamma}_h \| \le r_1 = \min\left(r, \; \beta C_1^2 \lambda_0 \right)$.

Theorems 1 and 2 remain valid if u is a vector valued function. Only the value of N changes: if u is a k-tupled vector, then $N = ks(2n+1)$. There are no changes in the formulation of the theorems and their proofs.

§3. The Stability of Variational-Difference Processes - Multidimensional Problems

Consider the equation

$$\sum_{|\alpha| = |\beta| = 0}^{s} (-1)^{\alpha} D^{\alpha} \left(A_{\alpha\beta} D^{\beta} u \right) = f(x) , \quad f \in L_2(\Omega) , \quad \Omega \subset \mathbb{R}_m . \tag{1}$$

Let $u(x)$ denote a k-tupled vector valued function with $k \ge 1$, and certain boundary conditions be associated with (1). As before we assume that the operator A induced by (1) and the boundary condition is positive definite in $L_2(\Omega)$. We study the stability of the variational-difference process for two independent situations simultaneously:

(1) the first boundary value problem for (1);

(2) arbitrary boundary conditions with Ω a parallelepiped.

In the latter, a linear transform of variables allows the parallelepiped to be mapped onto a cube, so we assume without loss of generality that Ω is the cube $\underline{0} \le x \le \underline{1}$.

We apply the variational-difference method to the problems under consideration. To be more precise, we consider an m-dimensional primitive system $\{\omega_q(t)\}$ with narrow support and degree $s - 1$, satisfying the fundamental completeness conditions (1.11) of Chapter II. Other choices of primitive systems yield analogous results.

The fundamental completeness conditions imply that the coordinate system

$$\varphi_{qjh}(x) = \omega_q\left(\frac{x}{h} - j\right) \tag{2}$$

is complete in H_A if (1) is non-degenerate. If it is degenerate, we assume in addition that (2) is complete in H_A.

If the first boundary value problem is considered, then, as usual, the approximate solution is sought in the form

$$u_h(x) = \sum_{|q|=0}^{s-1} \sum_{j\in\hat{\mathcal{J}}^h} a_{qj}^{(h)} \varphi_{qjh}(x) \tag{3}$$

and the coefficients $a_{qj}^{(h)}$ are determined by

$$\sum_{|q'|=0}^{s-1} \sum_{j\in\hat{\mathcal{J}}^h} a_{q'j'}^{(h)} [\varphi_{q'j'h}, \varphi_{qjh}]_A = (f, \varphi_{qjh}) \; , \quad |q| \le s-1 \; , \quad j \in \hat{\mathcal{J}}^h \; . \tag{4}$$

If arbitrary boundary conditions are considered on the cube $\underline{0} \le x \le \underline{1}$, then we put $h = 1/2n$, where n is a positive integer, and replace $\hat{\mathcal{J}}^h$ by $\mathcal{J}^h$ with

$$\mathcal{J}^h = \{j : -\underline{1} \le j \le 2\underline{n}-\underline{1}\} \; .$$

In both situations, the smaller cubes of the domains $\hat{\Omega}^h$ and Ω, respectively, which intersect Ω, lie completely in Ω.

In this section, for convenience of presentation, we introduce the notation Ω^h and $\mathcal{J}^h$ for the first boundary problem. Let $\mathcal{J}_0^h$ be the set of labels of the lower vertices of all the smaller cubes contained in Ω^h. We have with $Q_{j_0} = \{x : j_0 h \le x \le (j_0+1)h\}$ and $Q = \{x : \underline{0} \le x \le \underline{1}\}$

$$\|u_h\|^2 = \sum_{j_0 \in \mathcal{J}_0^h} \int_{Q_{j_0}} |u_h(x)|^2 dx = h^m \sum_{j_0 \in \mathcal{J}_0^h} \int_Q \left[\sum_{|q|=0}^{s-1} \sum_{i \in I} a^{(h)}_{q,j_0-i} \omega_q(t+i) \right]^2 dt \,. \tag{5}$$

It can be easily seen that the integral in the last term of (5) is a positive quadratic form in the variables $a^{(h)}_{q,j_0-i}$. Obviously, the coefficients of the form do not depend on h. Hence, it follows that there exist positive constants C_1 and C_2' such that

$$C_1^2 \sum_{|q|=0}^{s-1} \sum_{i \in I} \left[a^{(h)}_{q,j_0-i} \right]^2 \le \int_Q \left[\sum_{|q|=0}^{s-1} \sum_{i \in I} a^{(h)}_{q,j_0-i} \omega_q(t+i) \right]^2 dt$$

$$\le C_2'^2 \sum_{|q|=0}^{s-1} \sum_{i \in I} \left[a^{(h)}_{q,j_0-i} \right]^2 .$$

Multiplying this by h^m and taking the summation over $j_0 \in \mathcal{J}_0^h$, we obtain

$$C_1^2 h^m \sum_{j_0 \in \mathcal{J}_0^h} \sum_{|q|=0}^{s-1} \sum_{i \in I} \left[a^{(h)}_{q,j_0-i} \right]^2 \le \|u_h\|^2 \le C_2'^2 \sum_{j_0 \in \mathcal{J}_0^h} \sum_{|q|=0} \sum_{i \in I} \left[a^{(h)}_{q,j_0-i} \right]^2 .$$

We set $j_0 - i = j$. Obviously, $a^{(h)}_{qj} = 0$, if $j \notin \mathcal{J}^h$. Remembering this, we consider $j \in \mathcal{J}^h$. Each vector j enters the sum at least once and not more than 2^m-times. We bound the left hand side from below by the sum with each index counted only once. We bound the right hand side from above by the sum with each index j counted 2^m-times. As a result, we obtain

$$C_1 h^{m/2} \|a^{(h)}\|_{\mathbb{R}_N} \le \|u_h\| \le C_2 h^{m/2} \|a^{(h)}\|_{\mathbb{R}_N} \,, \quad C_2 = 2^{m/2} C_2' \,. \tag{6}$$

The notation in (6) is analogous to that of the preceding section: $a^{(h)}_{qj}$ is a k-tupled vector $a^{(h)}_{qj} = \left(a^{(h)}_{qj_1}, a^{(h)}_{qj_2}, \ldots, a^{(h)}_{qj_k} \right)$. We denote by $a^{(h)}$ the vector whose components are the coefficients $a^{(h)}_{qj_l}$, $0 \le |q| \le s-1$, $j \in \mathcal{J}^h$, $1 \le l \le k$, ordered in some appropriate way. If N denotes the number of these components, then $a^{(h)}$ is a vector in $\mathbb{R}_N$.

We denote by $F_{qj}^{(h)} = \left(f, \varphi_{qjh}\right)$, the components of the right hand side of (4). We assume that $f(x)$ has been extended by zero outside Ω. Then

$$F_{qj}^{(h)} = \int_{\Omega} f(x)\varphi_{qjh}(x)dx = \int_{\hat{Q}_j} f(x)\omega_q\left(\frac{x}{h} - j\right)dx ,$$

where $\hat{Q}_j$ is the grid cube in which $\varphi_{qjh}(x)$ is not identically equal to zero: $\hat{Q}_j = \left\{x : \underline{0} \le \frac{x}{h} - j \le \underline{2}\right\}$. Applying the Bunyakovsky-Schwartz inequality, we obtain

$$\left|F_{qj}^{(h)}\right| \le \int_{\hat{Q}_j} |f(x)|^2 dx \int_{\hat{Q}_j} \omega_q^2\left(\frac{x}{h} - j\right)dx .$$

The substitution $t = \frac{x}{h} - j$ in the second integral yields

$$\int_{\hat{Q}_j} \omega_q^2\left(\frac{x}{h} - j\right)dx = \gamma_q^2 h^m , \quad \gamma_q^2 = \int_{\underline{0}\le t\le \underline{2}} \omega_q^2(t)dt .$$

Hence

$$\left|F_{qj}^{(h)}\right|^2 \le \gamma_q^2 h^m \int_{\hat{Q}_j} |f(x)|^2 dx . \tag{7}$$

The set of vectors $F_{qj}^{(h)}$, $0 \le |q| \le s-1$, $j \in J^h$, can be examined as a vector $F^{(h)}$ in $\mathbb{R}_N$. Remembering this and summing the inequalities (7) over q and j, we obtain the estimate

$$\|F^{(h)}\|_{\mathbb{R}_N}^2 \le \gamma^2 h^m \|f\|_{L_2(\Omega^h)}^2 \le \gamma^2 h^m \|f\|_{L_2(\Omega)}^2 = \gamma^2 h^m \|f\|^2 , \quad \gamma^2 = 2^m \sum_{|q|=0}^{s-1} \gamma_q^2 . \tag{8}$$

We now introduce the N-dimensional Hilbert spaces X_N and Y_N with norms

$$\|a\|_{X_N} = h^{m/2}\|a\|_{\mathbb{R}_N} , \quad \|a\|_{Y_N} = h^{-m/2}\|a\|_{\mathbb{R}_N} . \tag{9}$$

These spaces are conjugate with respect to the inner product in $\mathbb{R}_N$. The inequalities (6) and (8) can now be rewritten in the form

$$C_1\|a^{(h)}\|_{X_N} \le \|u_h\| \le C_2\|a^{(h)}\|_{X_N} \tag{10}$$

and

$$\|F^{(h)}\|_{Y_N} \le \gamma\|f\| \tag{11}$$

As in the preceding section, we denote by M_h the matrix of (4), and by A_h the corresponding operator mapping X_N onto Y_N. The exact vector $a^{(h)}$ satisfies the equation

$$A_h a^{(h)} = F^{(h)} . \tag{12}$$

After repeating without change, the argument of §2, we obtain

$$\left\|A_h^{-1}\right\| \le \frac{1}{\lambda_0 C_1^2} , \tag{13}$$

where λ_0 is the infinium for the operator A. In the same way, we find that

$$\|a^{(h)}\|_{X_N} \le C = \text{const.} \tag{14}$$

Thus, by the corollary of §1, the following theorem is valid:

THEOREM 1. *The variation-difference process for the evaluation of the coefficients* $a_{qj}^{(h)}$ *of* (4) *is stable with respect to the sequence* (X_N, Y_N) .

In the same manner as in §2, the following theorem can be proved.

THEOREM 2. *The numerical process for the determination of the approximate solution* (3) *by the variational-difference method is stable with respect to the sequence of spaces* $\left\{H_A^{(N)}, Y_N\right\}$ *, where* $H_A^{(N)}$ *is the subspace of the space* H_A *spanned by functions of the form* (3).

Theorem 2 guarantees the existence of positive constants p_1, q_1 and r_1 such that, for $\|\tilde{\Gamma}_h\| < r_1$, the following inequality holds:

$$|\!|\!| \vartheta_h - u_h |\!|\!|_A \le p_1 \|\tilde{\Gamma}_h\| + q_1 \|\delta^{(h)}\|_{Y_N} . \tag{15}$$

The stability theorems enable an error estimate to be given for the approximate solution in terms of the error arising in the evaluation of the matrix and the right hand side of the variational-difference (algebraic) system. With such an aim in view, consider the methods introduced in §1, 2 of Chapter VIII. In §5 of Chapter VIII, we obtained that, if the components of the vector $F^{(h)}$ are calculated by the formula (1.13) of Chapter VIII, then the error-vector $\delta^{(h)}$ satisfies the inequality (1.15) of Chapter VIII. Consequently,

$$\|\delta^{(h)}\|_{Y_N} = h^{-m/2} \|\delta^{(h)}\|_{R_N} = O\left(h^{\bar{s}}\right) . \tag{16}$$

Further, if the elements of the matrix M_h are calculated with error $O\left(h^{\sigma+m}\right)$ then the norm of the operator $\tilde{\Gamma}_h$, which is induced by the error-matrix Γ_h and is considered as an operator in $\mathbb{R}_N$, is dominated by $O\left(h^{\sigma-2s+m/2}\right)$ according to formula (2.6) of Chapter VIII. It is not difficult to derive an estimate of the norm of this operator considered as an operator mapping X_N onto Y_N:

$$\|\tilde{\Gamma}_h\|_{X_N \to Y_N} = \sup_{a \in \mathbb{R}_N} \frac{\|\tilde{\Gamma}_h a\|_{Y_N}}{\|a\|_{X_N}} = h^{-m} \sup_{a \in \mathbb{R}_N} \frac{\|\Gamma_h a\|_{\mathbb{R}_N}}{\|a\|_{\mathbb{R}_N}} = O\left(h^{\sigma-2s-m/2}\right) . \tag{17}$$

It follows from (15) that

$$\|\vartheta_h - u_h\| \le p_2 h^{\sigma-2s-m/2} + q_2 h^{\bar{s}} , \tag{18}$$

where p_2 and q_2 are constants. The inequality (18) holds, provided $\|\tilde{\Gamma}_h\|_{X_N \to Y_N} \le r_1$. Thus, the error arising from the replacement of $u_h(x)$ by $\vartheta_h(x)$ is small for small values of h, if $\sigma > 2s + \frac{m}{2}$ and $s > 0$. If it is known that $\|u_h - u_0\|_{A} = O\left(h^{\mu}\right)$, $\mu > 0$, and we require that the replacement of u_h by ϑ_h does not degarde the last estimate, it is sufficient to require that

$$\bar{s} \ge \mu , \quad \sigma \ge 2s + \frac{m}{2} + \mu . \tag{19}$$

§4. The Stability of Variational-Difference Processes - Eigenvalue Problems

Let A be the operator considered in the preceding section. We assume in addition that its spectrum is discrete. Let $0 < \lambda_1 \le \lambda_2 \le \ldots \le \lambda_k \le \ldots$ denote its eigenvalues. We introduce the primitive system, coordinate functions and subspaces $H_A^{(N)}$ in the same way as in §3. Implementing the Rayleigh-Ritz method, one can find approximate eigenvalues $\lambda_1^{(N)}, \lambda_2^{(N)}, \ldots, \lambda_N^{(N)}$ as successive conditional extremums of the ratio

$$\frac{[\![\vartheta]\!]_A^2}{\|\vartheta\|^2} , \quad \vartheta \in H_A^{(N)} .$$

Assuming that the coordinate system is complete in H, we have, for fixed ,

$\lim_{N\to\infty} \lambda_k^{(N)} = \lambda_k$. As before, we denote by M_h the matrix of energy inner products $[\varphi_{q'j'h}, \varphi_{qjh}]_A$, and by S_h the matrix of inner products $(\varphi_{q'j'h}, \varphi_{qjh})$. If

$$\vartheta_h(x) = \sum_{|q|=0}^{s-1} \sum_{j \in J^h} b_{qj} \varphi_{qjh}(x) \tag{1}$$

and b is the vector with components b_{qj} , then

$$\|\vartheta\|_A^2 = (M_h b,\, b) \ , \quad \|\vartheta\|^2 = (S_h b,\, b) \ , \tag{2}$$

and $\lambda_k^{(N)}$, $k = 1, 2, \ldots, N$, are the eigenvalues of

$$(M_h - \lambda S_h)b = 0 \ . \tag{3}$$

We check whether the conditions of §1, and especially, the conditions of Theorem 2, §1, hold. In the present context $H_n = \mathbb{R}_N$, $A_n = M_h$, $B_n = S_h$. The operator $B_n^{\frac{1}{2}} A_n^{-\frac{1}{2}}$ is obviously completely continuous (compact) in H_n , since the space $\mathbb{R}_N$ is finite-dimensional. The third condition in (1.7) now takes the form $\|\vartheta\|^2 \le C_3^2 \|\vartheta\|_A^2$ where ϑ is an arbitrary function of the form (1). In the present context, this inequality is the consequence of the positive definiteness of the operator A . The second condition in (1.7) follows from the convergence of the approximate eigenvalues to the exact. It remains to verify the first condition in (1.7). In the next section, we shall prove the inequality $\mu_1^{(N)} \ge h^{-m}\lambda_1^{(N)}/C_2'$, where C_2' is a constant and $\mu_1^{(N)}$ is the smallest eigenvalue of the matrix M_h . Since $\lambda_1^{(N)} \xrightarrow[N\to\infty]{} \lambda_1$, we have that, for a sufficiently large N , $\mu_1^{(N)} \ge h^{-m}\lambda_1/2C_2' = \overline{C}h^{-m}$, $\overline{C} = \text{const.}$ Now $\left\|M_h^{-1}\right\| = 1/\mu_1^{(n)} \le h^m/\overline{C}$ which is bounded for bounded h . Finally, Theorem 2 of §1 yields that the variational-difference process is stable with respect to the sequence of spaces $\{\mathbb{R}_N\}$.

§5. On the Condition Number of the Variational-Difference Matrix*

We consider equation (3.1) under the conditions of §3, apply the variational-difference method considered in §3 to it and give an estimate of the condition number of the corresponding variational-difference matrix. The notations of §3 is retained. The condition number of the matrix M_h is

$$\kappa_N = \left(\mu_N^{(N)}\right) \Big/ \left(\mu_1^{(N)}\right) \ . \tag{1}$$

where

$$\mu_1^{(N)} = \min\left(\left\| \sum_{|q|=0}^{s-1} \sum_{j\in J^h} b_{qj}\varphi_{qjh}\right\|_A^2\right) \Big/ \left(\sum_{|q|=0}^{s-1} \sum_{j\in J^h} b_{qj}^2\right) , \tag{2a}$$

* See Mihklin [15].

$$\mu_N^{(N)} = \max\left(\left\| \sum_{|q|=0}^{s-1} \sum_{j\in J^h} b_{qj}\varphi_{qjh}\right\|_A^2\right)\Bigg/\left(\sum_{|q|=0}^{s-1} \sum_{j\in J^h} b_{qj}^2\right), \tag{2b}$$

where "min" and "max" are taken over the set of coefficients b_{qj} which are not all equal to zero.

We denote by $b_{qj}^{(0)}$ and $b_{qj}^{(1)}$ the values of b_{qj} for which the minimum and maximum in (2) are attained respectively. Let

$$w_0 = \sum_{|q|=0}^{s-1} \sum_{j\in J^h} b_{qj}^{(0)}\varphi_{qjh}, \quad w_1 = \sum_{|q|=0}^{s-1} \sum_{j\in J^h} b_{qj}^{(1)}\varphi_{qjh}, \tag{3}$$

so that

$$\mu_1^{(N)} = \left(\|w_0\|_A^2\right)\Bigg/\left(\sum_{|q|=0}^{s-1} \sum_{j\in J^h} \left[b_{qj}^{(0)}\right]^2\right), \quad \mu_N^{(N)} - \left(\|w_1\|_A^2\right)\Bigg/\left(\sum_{|q|=0}^{s-1} \sum_{j\in J^h} \left[b_{qj}^{(1)}\right]^2\right). \tag{4}$$

We can rewrite (3.6) as

$$C_1'\|u\|^2 \le h^m \sum_{|q|=0}^{s-1} \sum_{j\in J^h} a_{qj}^2 \le C_2'\|u\|^2, \quad C_1', C_2' = \text{const}, \tag{5}$$

where u is an arbitrary function of the form

$$u(x) = \sum_{|q|=0}^{s-1} \sum_{j\in J^h} a_{qj}\varphi_{qjh}(x). \tag{6}$$

From (5) the following inequalities are obtain

$$\begin{aligned} \mu_1^{(N)} &\ge \frac{h^{-m}}{C_2'}\frac{\|w_0\|_A^2}{\|w_0\|^2} \ge \frac{h^{-m}}{C_2'} \min_{u\in H_A^{(N)}} \frac{\|u\|_A^2}{\|u\|^2} = \frac{h^{-m}}{C_2'}\lambda_1^{(N)}, \\ \mu_N^{(N)} &\le \frac{h^{-m}}{C_1'}\frac{\|w_1\|_A^2}{\|w_1\|^2} \le \frac{h^{-m}}{C_1'} \max_{u\in H_A^{(N)}} \frac{\|u\|_A^2}{\|u\|} = \frac{h^{-m}}{C_1'}\lambda_N^{(N)}, \end{aligned} \tag{7}$$

where $\lambda_1^{(N)}$ and $\lambda_N^{(N)}$ are the variational-difference approximations of λ_1 and λ_N, respectively. We assume that the variational-difference process for the approximate eigenvalues converges, so $\lambda_1^{(N)} \ge C_3$, where C_3 is a non-zero constant. Then, as it was proved in §4 of Chapter VII, $\lambda_N^{(N)} \le C_4 h^{-2s}$, C_4 = const. Now, from (6) and (7), we obtain

$$\mu_1^{(N)} \geq C_5 h^{-m} \ , \quad \mu_N^{(N)} \leq C_6 h^{-m-2s} \ , \tag{8}$$

and therefore

$$\kappa_N \leq C_0 h^{-2s} \ , \quad C_0 = \text{const.} \tag{9}$$

We now prove that, for sufficiently large N, the following inequality holds:

$$\kappa_N \geq C_1 h^{-2s} \ , \quad C_1 = \text{const.} \tag{10}$$

Let u be an arbitrary function of the form (6). By (5) we have

$$\left(\|u\|_A^2\right) \Bigg/ \left(\sum_{|q|=0}^{s-1} \sum_{j \in J^h} a_{qj}^2\right) \leq \frac{h^{-m}}{C_1'} \frac{\|u\|_A^2}{\|u\|^2} \ .$$

We choose $u(x)$ so that the right-hand side of the last inequality takes its minimum value, which equals $h^{-m}\lambda_1^{(N)}/C_1'$. We replace the ratio on the left hand side by its smallest value $\mu_1^{(N)}$. As a result we obtain $\mu_1^{(N)} \leq h^{-m}\lambda_1^{(N)}/C_1'$. In the same way we find $\mu_N^{(N)} \geq h^{-m}\lambda_N^{(N)}/C_2'$ and hence

$$\kappa_N = \frac{\mu_N^{(N)}}{\mu_1^{(N)}} \geq \frac{C_1'}{C_2'} \cdot \frac{\lambda_N^{(N)}}{\lambda_1^{(N)}} \ .$$

But $\lambda_N^{(N)} \geq Ch^{-m}$, C = const., and $\lambda_1^{(N)} \xrightarrow[N\to\infty]{} \lambda_1$, so, if N is sufficiently large, then $\lambda_1^{(N)} \leq 2\lambda_1$. This yields the inequality (10). From (9) and (10) we obtain

$$\kappa_N \sim h^{-2s} \ . \tag{11}$$

We note that this estimate does not depend on the dimension of the space.

§6. The Case of Arbitrary Domains and Arbitrary Boundary Conditions

In the present context, the condition number of the variational-difference matrix can be arbitrary large. Below we present briefly the cunning work of Rukhovetz [1] who has proved the following: under certain regularity assumptions, one can transform the variational-difference process in such a way that the order of approximation does not change, but the condition number has the estimate $O\left(h^{-2s}\right)$ as above.

Let the boundary of the domain $\Omega \subset \mathbb{R}_m$ be sufficiently smooth. Consider the problem

$$(Au)(x) = f(x) \ , \quad f \in L_2(\Omega) \ , \tag{1}$$

where A is defined by the differential equation

$$Au = \sum_{|\alpha|=|\beta|=0}^{s} (-1)^{\alpha} D^{\alpha}\left(A_{\alpha\beta} D^{\beta} u\right) , \quad A_{\alpha\beta} = A^{*}_{\alpha\beta} , \tag{2}$$

along with appropriate natural boundary conditions. We assume that A is positive definite in $L_2(\Omega)$. In addition, we assume that

$$\|u\|_{2,s} \le CA(u, u) , \quad \forall u \in W_2^{(s)}(\Omega) , \tag{3}$$

where

$$A(u, \vartheta) = \int_{\Omega} \sum_{|\alpha|=|\beta|=0}^{s} A_{\alpha\beta} u^{(\alpha)} \vartheta^{(\beta)} dx . \tag{4}$$

Below we denote by C constants which differ in different situations.

We construct the grid with grid spacing h, and denote by $\check{\Omega}^h$ the subgrid consisting of the smaller mesh cubes intersecting Ω. Obviously, $\check{\Omega}^h \supset \Omega$. We let $\Omega_{(h)} = \check{\Omega}^h \backslash \Omega$, and introduce the functional

$$b_h(u, \vartheta) = \int_{\Omega_{(h)}} \sum_{|\alpha|=s} u^{(\alpha)} \vartheta^{(\alpha)} dx . \tag{5}$$

Following Rukhovetz [1], we replace the usual variational-difference (algebraic) system by

$$A\left(u_h, \varphi_{qjh}\right) + b_h\left(u_h, \varphi_{qjh}\right) = \left(f, \varphi_{qjh}\right) , \quad 0 \le |q| \le s-1 , \quad j \in \check{J}^h , \tag{6}$$

where $\check{J}^h$ is the set of labels of the lower vertices of the smaller mesh cubes contained in $\check{\Omega}^h$ and

$$u_h(x) = \sum_{|q|=0}^{s-1} \sum_{j \in \check{J}^h} a_{qj}^{(h)} \varphi_{qjh}(x) , \quad \varphi_{qjh}(x) = \omega_q\left(\frac{x}{h} - j\right) . \tag{7}$$

It is proved that the condition number of the variational-difference system (6) is $O\left(h^{-2s}\right)$. The approximations (7) has the same order of approximation as the approximate solutions obtained by the usual variational-difference schemes. The paper of Rukhovetz [1] lists the conditions on the primitive system which guarantee the last mentioned results. They hold for product systems and systems with wide support.

§7. Numerical Example - A Degenerate Second Order Ordinary Differential Equation

1. Consider the differential equation

$$-\frac{d}{dx}\left(x^{\alpha}\frac{du}{dx}\right) + u = \frac{x^{3-\alpha}-2(3-\alpha)x-1}{3-\alpha} \tag{1}$$

on the interval $0 < x < 1$. Since we assume that $1 \leq \alpha \leq 2$, it is sufficient to prescribe only the boundary condition at $x = 1$. Let

$$u(1) = 0 \ . \tag{2}$$

The exact solution is

$$u = \frac{x^{3-\alpha}-1}{3-\alpha} \ . \tag{2}$$

We shall construct an approximate solution by the variational-difference method. Following Chapter V, we use the primitive function

$$\omega(t) = \begin{cases} t \ , & 0 \leq t \leq 1 \ , \\ 2-t \ , & 1 < t \leq 2 \ , \\ 0 \ , & \end{cases} \tag{3}$$

construct the coordinate functions

$$\varphi_{jh}(x) = \omega\left(\frac{x}{h} - j\right)$$

and seek the approximate solution in the form

$$u_h(x) = \sum_{j=-1}^{2n-2} a_j\varphi_{jh}(x) \ . \tag{4}$$

Our problem is the particular case of the problem examined in Chapter V when $p(x) = 1$, $q(x) = 1$. Consequently, the inequality $q(x) \geq q_0 = \text{const.} > 0$, mentioned several times in Chapter V, holds with $q_0 = 1$.

Since

$$f(x) = \frac{x^{3-\alpha}-2(3-\alpha)x-1}{3-\alpha} \tag{5}$$

it follows that $f \in L_\infty(0, 1)$. Using the estimate (6.4) of Chapter V, we have

$$\|u-u_h\| = O\left(h^{(3-\alpha)/2}\right) \ , \tag{6}$$

where, in the present context,

$$\|u\|^2 = \int_0^1 \left[x^{\alpha}\left(\frac{du}{dx}\right)^2 + u^2\right] dx \ . \tag{7}$$

From (6) it follows that, for any interval $[\delta, 1]$, where $0 < \delta < 1$, the following uniform estimate holds:

$$|u(x)-u_h(x)| = O\left(h^{(3-\alpha)/2}\right) \ . \tag{8}$$

Indeed, by the boundary condition (2),

$$u(x) - u_h(x) = -\int_x^1 [u'(t)-u_h'(t)]dt = -\int_x^1 t^{\alpha/2}[u'(t)-u_h'(t)]\frac{dt}{t^{\alpha/2}}$$

and by Bunyakovsky-Schwartz inequality

$$|u(x)-u_h(x)| \le \|u-u_h\|\left\{\int_x^1 \frac{dt}{t^\alpha}\right\}^{\frac{1}{2}} = \begin{cases} \sqrt{\left(1-x^{\alpha-1}\right)/(\alpha-1)}\ \|u-u_h\| , & \alpha > 1 , \\ \sqrt{\ln 1/x}\ \|u-u_h\| , & \alpha = 1 . \end{cases} \tag{9}$$

The variational-difference system from which the coefficients a_j are obtained is therefore

$$\sum_{k=-1}^{2n-1} [\varphi_{kh}, \varphi_{jh}]a_k = (f, \varphi_{jh}) , \quad -1 \le j \le 2n-2 . \tag{10}$$

2. We now examine the calculation of the right hand side of (10). We write

$$f^{(j)} = (f, \varphi_{jh}) = \int_0^1 f(x)\omega\left(\frac{x}{h} - j\right)dx . \tag{11}$$

If $j = -1$, then

$$f^{(-1)} = \int_0^h f(x)\omega\left(\frac{x}{h} + 1\right)dx = \int_0^h f(x)\left(1 - \frac{x}{h}\right)dx = \frac{1}{3-\alpha}\left[\frac{h^{4-\alpha}}{(4-\alpha)(5-\alpha)} - \frac{3-\alpha}{3} - \frac{h}{2}\right] . \tag{12}$$

If $0 \le j \le 2n-2$, then we must in fact integrate over an interval of length $2h$; namely

$$f^{(j)} = \int_{jh}^{(j+2)h} f(x)\omega\left(\frac{x}{h} - j\right)dx .$$

A change of variables $x = (j+t)h$ gives

$$f^{(j)} = h\int_0^2 f\big((j+t)h\big)\omega(t)dt = \frac{h^{4-\alpha}[(j+2)^{5-\alpha}-2(j+1)^{5-\alpha}+j^{5-\alpha}]}{(3-\alpha)(4-\alpha)(5-\alpha)} - \frac{h}{3-\alpha}[2(3-\alpha)(j+1)h+1] .$$

We can simplify the last expression by putting $jh = x_j$

$$f^{(j)} = \frac{\left(x_j+2h\right)^{5-\alpha}-2\left(x_j+h\right)^{5-\alpha}+x_j^{5-\alpha}}{(3-\alpha)(4-\alpha)(5-\alpha)h} - 2x_{j+1}h - \frac{h}{3-\alpha} . \tag{13}$$

We note that, for sufficiently small h , the evaluation of (13) can become unstable. In this situation, we can expand $\left(x_j+2h\right)^{5-\alpha}$ and $\left(x_j+h\right)^{5-\alpha}$ using Taylor series, which thereby yields the following formula which is valid for suitably large values of x_j :

$$f^{(j)} = \frac{h}{3-\alpha} x_j^{3-\alpha} - 2(x_j+h)h - \frac{h}{3-\alpha} + h^2 x_j^{2-\alpha} + O(h^3) . \tag{14}$$

Obviously, the approximation (14) can be made more accurate, but at the expense of being more complicated.

3. Next, we turn to the calculation of the matrix of system (10). We distinguish three cases:

(a) the diagonal element $[\varphi_{-1,h}, \varphi_{-1,h}]$;

(b) the remaining diagonal elements;

(c) the off-diagonal elements.

We note that the diagonal elements are calculated by the formula

$$[\varphi_{jh}, \varphi_{jh}] = \int_0^1 \left[\frac{x^\alpha}{h^2} \omega'^2\left(\frac{x}{h} - j\right) + \omega^2\left(\frac{x}{h} - j\right)\right] dx . \tag{15}$$

(a) If $j = -1$, then

$$[\varphi_{-1,h}, \varphi_{-1,h}] = \int_0^h \left[\frac{x^\alpha}{h^2} + \left(1 - \frac{x}{h}\right)^2\right] dx = \frac{h^{\alpha-1}}{\alpha+1} + \frac{h}{3} . \tag{16}$$

(b) If $j \geq 0$, then

$$[\varphi_{jh}, \varphi_{jh}] = \int_{jh}^{(j+2)h} \frac{x^\alpha}{h^2} dx + \int_{jh}^{(j+2)h} \omega^2\left(\frac{x}{h} - j\right) dx .$$

The last integral can be easily evaluated and we obtain as a result

$$[\varphi_{jh}, \varphi_{jh}] = \frac{(x_j+2h)^{\alpha+1} - x_j^{\alpha+1}}{(\alpha+1)h^2} + \frac{2h}{3} , \quad 0 \leq j \leq 2n-2 . \tag{17}$$

For suitably large values of x_j , one can obtain the following formula

$$[\varphi_{jh}, \varphi_{jh}] = \frac{2x_j^\alpha}{h} + 2\alpha x_j^{\alpha-1} + \frac{2h}{3} + \frac{4}{3}\alpha(\alpha-1)x_j^{\alpha-2}h - \frac{2}{3}(2-\alpha)(\alpha-1)x_j^{\alpha-3}h^2 + O(h^3) . \tag{18}$$

(c) We turn to the evaluation of the off-diagonal elements. If $|k-j| > 1$, then the supports of functions $\varphi_j^{(h)}(x)$ and $\varphi_k^{(h)}(x)$ do not overlap and therefore $[\varphi_{kh}, \varphi_{jh}] = 0$. Therefore, the matrix (10) is tri-diagonal. Since it is also symmetric, it is sufficient to calculate elements of the form $[\varphi_{jh}, \varphi_{j+1,h}]$. The supports of the functions φ_{jh} and $\varphi_{j+1,h}$ are the intervals $[jh, (j+2)h]$ and $[(j+1)h, (j+3)h]$, respectively. They overlap on the interval $[(j+1)h, (j+2)h]$, so

$$[\varphi_{jh},\varphi_{j+1,h}] = \int_{(j+1)h}^{(j+2)h} \left[\frac{x^\alpha}{h^2}\,\omega'\left(\frac{x}{h} - j\right)\omega'\left(\frac{x}{h} - j - 1\right) + \omega\left(\frac{x}{h} - j\right)\omega\left(\frac{x}{h} - j - 1\right)\right]dx\ .$$

Using (3), we find

$$[\varphi_{jh},\varphi_{j+1,h}] = -\frac{1}{h^2}\int_{(j+1)h}^{(j+2)h} x^\alpha dx + \int_{(j+1)h}^{(j+2)h}\left(2 + j - \frac{x}{h}\right)\left(\frac{x}{h} - j - 1\right)dx$$

$$= \frac{h}{6} - \frac{h^{\alpha-1}}{\alpha+1}\left[(j+2)^{\alpha+1} - (j+1)^{\alpha+1}\right]\ . \quad (19)$$

The approximate formula for suitably large values of x_j has the form

$$[\varphi_{jh},\ \varphi_{j+1,h}] = \frac{x^\alpha_{j+1}}{h} - \frac{\alpha}{2}\,x^{\alpha-1}_{j+1} - \frac{(\alpha-1)\alpha}{6}\,x^{\alpha-2}_{j+1}h + \frac{h}{6} + \frac{(2-\alpha)(\alpha-1)\alpha}{24}\,x^{\alpha-3}_{j+1}h^2 + O(h^3)\ . \quad (20)$$

It is worth mentioning why we used the approximations (14), (18) and (20) with error of order $O(h^3)$. The maximum accuracy which is guaranteed by (6) is $O(h)$. The error arising from the mentioned approximations must be of the same or higher order. From (2.18), it is clear that the error of the non-exact approximate solution has the same order as the values of $\|\tilde{\Gamma}_h\|$ and $\|\delta^{(h)}\|_{Y_N}$. Since the remainder in (14) has error $O(h^3)$, $\|\delta^{(h)}\|_{\mathbb{R}_N} = O(h^{5/2})$ and hence $\|\delta^{(h)}\|_{Y_N} = O(h^2)$. Further, the error of each non-zero element in the matrix of (10) is $O(h^3)$. The number of such elements is $O(h^{-1})$, and the norm of the error matrix, as an operator from X_N to Y_N, has the estimate $O(h^{3/2})$. Thus, the chosen order for the error in the elements of the matrix and right hand sides is such that the error of the nonexact approximation has a higher order than the theoretical error of the approximate solution. If the order of the error in (14), (18) and (20) is lowered by one, then the order of the error of the nonexact approximations can become less than the order in (6).

All calculations relevant to this example, including the algorithm and programming in ALGOL-60 on a BESSM-6 computer were performed by K.G. Semjenova to whom the author is greatly indebted.

The variational-difference method was applied to equation (1) for the following values $\alpha =: 1,\ 1.5,\ 1.8,\ 2$. For $\alpha = 1$ the calculations were performed with $h =: 0.01,\ 0.001,\ 0.0001$. For the remaining values of α - with $h = 0.01$ and 0.001. For all values of α with $h = 0.01$ and $0.0.1$, the calculations were performed using the exact formulas (13), (17) and (19), as well as using the approximations (14), (18) and (20). We note that, for $\alpha = 1$ and $\alpha = 2$, the approximations (18) and (20) become exact. For $h = 0.0001$, only approximate formulas were used.

In all cases the variational-difference system was solved by the run method (a variant of the classical Gauss method, designed specifically for the solution of tri-diagonal systems). For further details about this method see, for example, the book of Godunov and Riabenkii [1].

Below, in Tables 3-5, we list absolute values of $u_h(x)$ minus the solution of the examined problem for $x = 0.0, 0.1, 0.2, \ldots, 0, 0.9$ and $\alpha = 1, 1.5, 1.8$. The tables show that, for $h = 0.01$, the exact formulas (13), (17) and (19) lead to more accurate results than the non-exact ones; and for $h = 0.001$ - to less accurate results. In addition, the accuracy of the solution itself for $h = 0.001$ is better than for $h = 0.01$. Thus, the conclusion is: when the grid size is decreased from 0.01 to 0.001, the growth of the condition number of the variational-difference system does not yet influence the results, whereas the lack of stability of the exact formulas (13), (17) and (19) does. For $h = 0.0001$, a noticeable increase in the error is apparent. Thus, the increase in the condition number of the variational-difference matrix appears to be an influence at this stage.

For $\alpha = 2$, the results of the calculations were practically exact. The reason is that, for $\alpha = 2$, $u(x) = x - 1$ which is linear. Therefore it can be easily established that $u_h(x) \equiv u(x)$.

The tables reveal that the error decreases as α increases from 1 to 2. It is most likely explained by the fact that as α increases, $u(x)$ becomes linear and the accuracy of the variational-difference approximation for $u(x)$ increases.

TABLE 3

Absolute Errors for $\alpha = 1$

x	$h = 0.01$		$h = 0.001$		$h = 0.0001$
	Exact Formulas	Nonexact Formulas	Exact Formulas	Nonexact Formulas	Nonexact Formulas
0.0	0.33×10^{-4}	0.51×10^{-4}	0.11×10^{-4}	0.10×10^{-5}	0.20×10^{-3}
0.1	0.15×10^{-4}	0.36×10^{-4}	0.13×10^{-4}	0.70×10^{-6}	0.18×10^{-3}
0.2	0.11×10^{-4}	0.32×10^{-4}	0.14×10^{-4}	0.61×10^{-6}	0.13×10^{-3}
0.3	0.83×10^{-5}	0.28×10^{-4}	0.15×10^{-4}	0.54×10^{-6}	0.87×10^{-4}
0.4	0.66×10^{-5}	0.25×10^{-4}	0.16×10^{-4}	0.47×10^{-6}	0.65×10^{-4}
0.5	0.51×10^{-5}	0.21×10^{-4}	0.14×10^{-4}	0.40×10^{-6}	0.40×10^{-4}
0.6	0.39×10^{-5}	0.17×10^{-4}	0.12×10^{-4}	0.32×10^{-6}	0.34×10^{-4}
0.7	0.29×10^{-5}	0.13×10^{-4}	0.98×10^{-5}	0.24×10^{-6}	0.23×10^{-4}
0.8	0.18×10^{-5}	0.86×10^{-5}	0.67×10^{-5}	0.16×10^{-6}	0.14×10^{-4}
0.9	0.86×10^{-6}	0.42×10^{-5}	0.33×10^{-5}	0.81×10^{-7}	0.67×10^{-5}

TABLE 4

Absolute errors for $\alpha = 1.5$

x	$h = 0.01$		$h = 0.001$	
	Exact formulas	Nonexact formulas	Exact formulas	Nonexact formulas
0.0	0.81×10^{-4}	0.24×10^{-3}	0.25×10^{-5}	0.49×10^{-5}
0.1	0.17×10^{-4}	0.25×10^{-3}	0.93×10^{-5}	0.31×10^{-5}
0.2	0.10×10^{-4}	0.20×10^{-3}	0.12×10^{-4}	0.24×10^{-5}
0.3	0.69×10^{-5}	0.14×10^{-3}	0.14×10^{-4}	0.17×10^{-5}
0.4	0.51×10^{-5}	0.10×10^{-3}	0.14×10^{-4}	0.12×10^{-5}
0.5	0.38×10^{-5}	0.74×10^{-4}	0.13×10^{-4}	0.88×10^{-6}
0.6	0.28×10^{-5}	0.52×10^{-4}	0.11×10^{-4}	0.63×10^{-6}
0.7	0.19×10^{5}	0.35×10^{-4}	0.85×10^{-5}	0.44×10^{-6}
0.8	0.16×10^{-5}	0.21×10^{-4}	0.57×10^{-5}	0.27×10^{-6}
0.9	0.53×10^{-6}	0.16×10^{-4}	0.28×10^{-5}	0.13×10^{-6}

TABLE 5

Absolute Errors for $\alpha = 1.8$

x	$h = 0.01$		$h = 0.001$	
	Exact formulas	Nonexact formulas	Exact formulas	Nonexact formulas
0.0	0.84×10^{-4}	0.13×10^{-3}	0.38×10^{-5}	0.63×10^{-5}
0.1	0.10×10^{-4}	0.13×10^{-3}	0.91×10^{-5}	0.16×10^{-5}
0.2	0.56×10^{-5}	0.11×10^{-3}	0.13×10^{-4}	0.12×10^{-5}
0.3	0.36×10^{-5}	0.72×10^{-4}	0.15×10^{-4}	0.82×10^{-6}
0.4	0.25×10^{-5}	0.49×10^{-4}	0.15×10^{-4}	0.57×10^{-6}
0.5	0.18×10^{-5}	0.34×10^{-4}	0.14×10^{-4}	0.41×10^{-6}
0.6	0.13×10^{-5}	0.24×10^{-4}	0.12×10^{-4}	0.29×10^{-6}
0.7	0.87×10^{-6}	0.16×10^{-4}	0.93×10^{-5}	0.20×10^{-6}
0.8	0.53×10^{-6}	0.92×10^{-5}	0.62×10^{-5}	0.13×10^{-6}
0.9	0.23×10^{-6}	0.40×10^{-5}	0.30×10^{-5}	0.59×10^{-7}

CHAPTER X

THE EULER-MACLAURIN SUM FORMULA

§1. A New Derivation of the Euler-Maclaurin Sum Formula

It is obvious how approximations for given functions, of one or several real variables, can be used to generate quadrature or cubature formulas. We show in the present section how the Euler-Maclaurin sum formula can be derived from the variational-difference approximation (5.1) of Chapter I, when applied to a function of one independent variable (see Mikhlin [18]).

Let x denote a real variable in $\mathbb{R}$ and let $u(x)$ belong to $W_p^{(s+1)}(0, 1)$, where s is an integer and p is any real number such that $1 \le p \le \infty$. In addition, let $\{\omega_q(t)\}$ denote a one-dimensional primitive system, with narrow support and degree $s - 1$, satisfying the fundamental completeness conditions, which in the present context take the form (6.1) and (6.2) of Chapter II. Putting

$$u^h(x) = \sum_{q=0}^{s-1} \sum_{j=-1}^{2n-1} h^q u^{(q)}((j+1)h)\omega_q\left(\frac{x}{h} - j\right) = \sum_{q=0}^{s-1} \sum_{j=0}^{2n} h^q u^{(q)}(jh)\omega_q\left(\frac{x}{h} - j + 1\right), \quad (1)$$

we obtain from §3 of Chapter III that

$$\|u-u^h\|_{p,\bar{s}} \le C\|u\|_{p,s+1}h^{s-\bar{s}+1}, \quad 0 \le \bar{s} \le s \; ; \quad C = \text{const.} \quad (2)$$

Hence, the following quadrature formula is derived

$$\int_0^1 u(x)dx = \sum_{q=0}^{s-1} \sum_{j=0}^{2n} h^q u^{(q)}(jh) \int_0^1 \omega_q\left(\frac{x}{h} - j + 1\right)dx + \rho_h, \quad (3)$$

where, for $u \in W_p^{(s+1)}$, the remainder ρ_h can be estimated as

$$|\rho_h| \le C\|u\|_{p,s+1} h^{s+1} . \tag{4}$$

The support of the function $\omega_q\left(\frac{x}{h} - j + 1\right)$ lies within the segment $(j-1)h \le x \le (j+1)h$, so when calculating the integrals in (3) it is necessary to distinguish between three situations:

(1) $j = 0$;

(2) $j = 2n$;

(3) $1 \le j \le 2n-1$.

In the first, we have

$$\int_0^1 \omega_q\left(\frac{x}{h} + 1\right)dx = \int_0^h \omega_q\left(\frac{x}{h} + 1\right)dx = h\int_1^2 \omega_q(t)dt \overset{\text{def}}{=\!=} ha_q .$$

In the second, we have

$$\int_0^1 \omega_q\left(\frac{x}{h} - 2n + 1\right)dx = \int_{(2n-1)h}^{2nh} \omega_q\left(\frac{x}{h} - 2n + 1\right)dx = h\int_0^1 \omega_q(t)dt \overset{\text{def}}{=\!=} hb_q .$$

Finally, if $1 \le j \le 2n-1$, then

$$\int_0^1 \omega_q\left(\frac{x}{h} - j + 1\right)dx = \int_{(j-1)h}^{(j+1)h} \omega_q\left(\frac{x}{h} - j + 1\right)dx = h\int_0^2 \omega_q(t)dt = h\left(a_q+b_q\right) .$$

Thus,

$$\int_0^1 u(x)dx = h\left\{\sum_{q=0}^{s-1}\sum_{j=1}^{2n-1} h^q u^{(q)}(jh)\left(a_q+b_q\right) + \sum_{q=0}^{s-1}\left[a_q u^{(q)}(0)+b_q u^{(q)}(1)\right]\right\} + \rho_h . \tag{5}$$

If $s = 1$, then $q = 0$, and the only primitive function is $\omega_0(t) = t$ for $0 \le t \le 1$ and $\omega_0(t) = 2 - t$ for $1 < t \le 2$. Here $a_0 = b_0 = \frac{1}{2}$ and (5) becomes the trapezoidal rule. Let $s > 1$. Then, it is known that the primitive functions $\omega_q(t)$ are not uniquely defined: for example, we can take arbitrary functions $\omega_q(t)$, $q \le s-2$, for $0 \le t \le 1$ and find the function $\omega_{s-1}(t)$ for $0 \le t \le 1$ and all the functions $\omega_q(t)$, $0 \le q \le s-1$ for $1 < t \le 2$ from (6.1) and (6.2) of Chapter II. We shall prove that the functions $\omega_q(t)$, $0 \le q \le s-2$, $0 \le t \le 1$, can be chosen so that the following conditions hold

$$a_0 = b_0 = \tfrac{1}{2} , \quad a_q + b_q = 0 , \quad q > 0 .$$

Choose functions $\bar{\omega}_q(t)$, $0 \le q \le s-2$, $0 \le t \le 1$, satisfying conditions consistent with the conditions (1)-(3) of §3, Chapter I. For $m = 1$, they take the form

$$\bar{\omega}_q \in W_p^{(s)}(0, 1) \cap C^{(s-1)}[0, 1] , \tag{7}$$

$$\bar{\omega}_q^{(\alpha)}(0) = 0 \ , \quad \bar{\omega}_q^{(\alpha)}(1) = \delta_{\alpha q} \ , \quad 0 \le \alpha, q \le s-2 \ . \tag{8}$$

The functions $\omega_q(t) = \bar{\omega}_q(t) + A_q t^s(1-t)^s$, with A_q = const., satisfy the same conditions, using the fundamental completeness conditions. We extend the functions $\omega_q(t)$, $0 \le q \le s-1$, to have compact support on the interval $0 \le t \le 2$. We denote by $\bar{a}_q$ and $\bar{b}_q$ the values of the constants a_q and b_q in $\bar{\omega}_q$. Obviously,

$$b_q = \bar{b}_q + \mu_s A_q \ , \quad \mu_s = \int_0^1 t^s(1-t)^s dt = \frac{(s!)^2}{(2s+1)!} \ . \tag{9}$$

We integrate the fundamental completeness conditions (6.1) and (6.2) of Chapter II over the interval $[0, 1]$:

$$\sum_{q=0}^{s-1} \frac{b_q}{(s-q)!} = \frac{1}{(s+1)!} \ ,$$

$$a_r + \sum_{q=0}^{r} \frac{b_q}{(r-q)!} = \frac{1}{(r+1)!} \ , \quad 0 \le r \le s-1 \ . \tag{10}$$

For $r = 0$, we find from (10) that $a_0 + b_0 = 1$. Then, for $r = 1$, we obtain $a_1 + b_1 + b_0 = \frac{1}{2}$. We require $b_0 = \frac{1}{2}$ or $\bar{b}_0 + A_0\mu_s = \frac{1}{2}$. Then A_0 is defined with $a_0 = b_0 = \frac{1}{2}$ and $a_1 + b_1 = 0$. For $r = 2$, we have

$$a_2 + b_2 + b_1 + \frac{1}{2} b_0 = a_2 + b_2 + \bar{b}_1 + A_1\mu_s + \frac{1}{4} = \frac{1}{6} \ .$$

We require $\bar{b}_1 + A_1\mu_s + 1/4 = 1/6$. This defines the value of A_1 for which $a_2 + b_2 = 0$. Proceeding in the same way we find the coefficients A_q , $q \le s-2$, and at the same time satisfy the relations (6) for $q \le s-1$. In addition, we determine the values of a_q and b_q for $q \le s-2$. Finally, from the first equation in (10), we find b_{s-1} , and then a_{s-1} from the equality $a_{s-1} + b_{s-1} = 0$. The relations (6) remain valid. Eliminating the a_r terms from (10), we obtain

$$\sum_{q=0}^{s-1} \frac{-b_q}{(r-q)!} = -\frac{1}{(r+1)!} \ , \quad r = 1, 2, \ldots \ . \tag{11}$$

These equations reveal that the coefficients b_q are independent of the choice of the primitive system and of the integer s .

Using (6), we can simplify the quadrature formula (5),

$$\int_0^1 u(x)dx = h\left[\sum_{j=1}^{2n-1} u(jh) + \frac{u(0)+u(1)}{2} + \sum_{q=1}^{s-1} b_q\left(u^{(q)}(0)-u^{(q)}(1)\right)\right] + \rho_h \ . \tag{12}$$

The coefficients can be easily calculated. Consider the function

$$f(z) = \frac{1}{e^z - 1} - \frac{1}{z} . \tag{13}$$

It is holomorphic near the point $z = 0$ and therefore can be expanded in a power series about $z = 0$,

$$f(z) = \sum_{k=0}^{\infty} \alpha_k z^k . \tag{14}$$

It is known (see, for example, Whittaker and Robinson [1], §67) that $\alpha_0 = -\frac{1}{2}$, $\alpha_{2k} = 0$, $k > 0$, $\alpha_{2k+1} = (-1)^k B_k/k!$ where the B_k denote the Bernoulli numbers. From (13) and (14), we find

$$\left(\frac{1}{z} + \sum_{n=0}^{\infty} \alpha_n z^n \right) \sum_{k=1}^{\infty} \frac{z^k}{k!} = 1 .$$

A comparison of coefficients yields relations identical with (11). Hence, it follows that $b_q = -\alpha_q$ and (12) coincides with the Euler-Maclaurin sum formula.

§2. A Related Euler-Maclaurin Sum Formula

Consider $u \in W_p^{(2s)}(0, 1)$ with $1 \le p \le \infty$ and s a positive integer. We introduce the primitive functions $\omega_q(t)$, $q = 0, 1, \ldots, s-1$, constructed as in §6 of Chapter II. They guarantee the highest order of approximation. Using them, we construct the function $\check{\omega}(t)$ defined by (6.10) of Chapter VIII, and the approximations $\check{u}^h(x)$ defined by (6.14) of Chapter VIII. Then

$$\|u - \check{u}^h\| \le C \|u\|_{p,2s} h^{2s} .$$

If we use the approximation

$$\int_0^1 u(x)dx \approx \int_0^1 \check{u}^h(x)dx \tag{1}$$

then the error associated with it will be of order $O\left(h^{2s}\right)$. By (6.14) of Chapter VIII,

$$\int_0^1 u^h(x)dx = \sum_{j=-\sigma-1}^{2n+\sigma-1} u\left((j+1)h\right) \int_0^1 \check{\omega}\left(\frac{x}{h} - j\right)dx = \sum_{j=-\sigma}^{2n+\sigma} g_j u(jh) ,$$

$$g_j = \int_0^1 \check{\omega}\left(\frac{x}{h} - j + 1\right)dx . \tag{2}$$

The coefficients g_j are calculated using (6.10) of Chapter VIII:

$$g_j = \sum_{q=0}^{s-1} \sum_{l=-\sigma}^{\sigma} b_{ql}^{(\sigma)} \int_0^1 \omega_q\left(\frac{x}{h} + l - j + 1\right)dx \ . \tag{3}$$

Consider the case when

$$\sigma+1 \le j \le 2n-\sigma-1 \ . \tag{4}$$

The support of the function under the integral sign in (3) is contained within the interval $[(j-l-1)h,\ (j-l+1)h] \subset [0,\ 1]$, so

$$\int_0^1 \omega_q\left(\frac{x}{h} + l - j + 1\right)dx = \int_{(j-l-1)h}^{(j-l+1)h} \omega_q\left(\frac{x}{h} + l - j + 1\right)dx = h^2 \int_0^2 \omega_q(t)dt \ . \tag{5}$$

We obtain, for the values of j satisfying (4),

$$g_j = h \sum_{q=0}^{s-1} \int_0^2 \omega_q(t)dt = \sum_{l=-\sigma}^{\sigma} b_{ql}^{(\sigma)} \ .$$

By (6.4) of Chapter VIII, the inner sum equals zero for $q \ne 0$. If $q = 0$, then (see §6 of Chapter VIII) $b_{00}^{(\sigma)} = 1$, $b_{0l}^{(\sigma)} = 0$, $l \ne 0$, and the inner sum equals unity. Thus

$$g_j = h \int_0^2 \omega_0(t)dt \ .$$

We put $r = 0$ in (6.2) of Chapter II and integrate it from zero to one. We obtain

$$\int_0^2 \omega_0(t)dt = 1 \ ,$$

and hence $g_j = h$, if j satisfies (4). We thereby obtain the following quadrature formula

$$\int_0^1 u(x)dx = \sum_{j=-\sigma}^{\sigma} g_j u(jh) + \sum_{j=2n-\sigma}^{2n+\sigma} g_j u(jh) + h \sum_{j=\sigma+1}^{2n-\sigma-1} u(jh) + \rho_h \tag{6}$$

where this time

$$|\rho_h| \le C\|u\|_{p,2s} h^{2s} \tag{7}$$

with the values of g_j given by (3). A similar quadrature formula can be obtained from (6.15) of Chapter VIII. For this, an extension, outside the interval of integration, is required for the functions under the integral sign in (6).

§3. An Euler-Maclaurin Sum Formula for the Multidimensional Cube

Let Q denote the m-dimensional cube $\underline{0} \le x \le \underline{1}$, and consider the evaluation of the integral

$$L = \int_Q u(x)dx \tag{1}$$

where $u \in W_p^{(s+1)}(Q)$. We retain the notation and formulas of §1 without change. We introduce the one-dimensional primitive system $\{\omega_q(x)\}$ considered in §1 and construct the m-dimensional product primitive system

$$\tilde{\omega}_q(x) = \prod_{\nu=1}^{m} \omega_{q_\nu}(x_\nu) , \quad |q| \le s . \tag{2}$$

Let the approximations be

$$u^h(x) = \sum_{|q|=0}^{s} \sum_{j=-1}^{2n-1} h^q u^{(q)}((j+1)h)\tilde{\omega}_q\left(\frac{x}{h} - j\right) . \tag{3}$$

As we know

$$\|u-u^h\|_p \le Ch^{s+1} , \tag{4}$$

where C does not depend on h. After integration, we obtain the cubature formula

$$\int_Q u(x)dx = \int_Q u^h(x)dx + \rho_h = \sum_{|q|=0}^{s} \sum_{j=-1}^{2n-1} h^q u^{(q)}((j+1)h) \int_Q \tilde{\omega}_q\left(\frac{x}{h} - j\right)dx + \rho_h ,$$

$$\rho_h = O(h^{s+1}) . \tag{5}$$

The support of the function $\tilde{\omega}_q\left(\frac{x}{h} - j\right)$ is the cube $\tilde{Q}_j \equiv \{x : jh \le x \le (j+2)h\}$. This is one of the larger cubes of the grid with lower vertex jh. As usual, we denote by J^h the set of vertices j over which the summation in (3) and (5) is taken: $J^h = \{j : -1 \le j \le 2n-1\}$. We set $J^h = J_A^h + J_B^h$, where J_A^h is the set of labels j which correspond to the larger grid-cubes which lie in Q. If $j \in J_B^h$, then only part of the cubes $\tilde{Q}_j$ lies in Q. More definitely, there exist integers k, $0 \le k \le m-1$, such that, of the set of 2^m small cubes which form $\tilde{Q}_j$, only 2^k of the small cubes lie in Q (the remainder lie outside). We denote by J_k^h the set of vertices $j \in J_B^h$ for which k has the same value. Hence, $J_B^h = \bigcup_{k=0}^{m-1} J_k^h$. For the sake of uniformity, we also write $J_A^h = J_m^h$.

Figure 17 depicts the case $m = 2$. The grid points with labels belonging to the

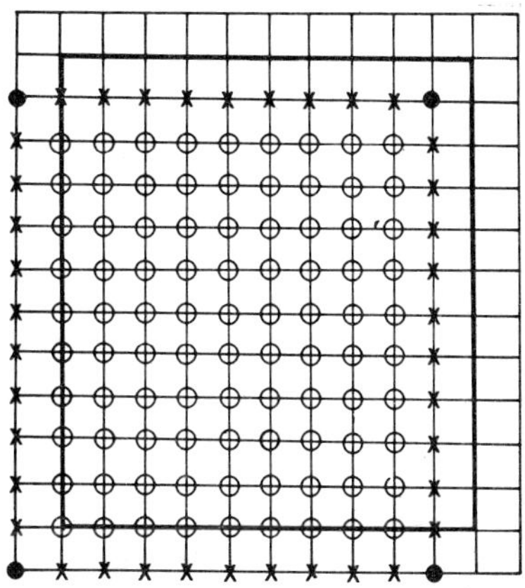

Fig. 17

sets J_0^h, J_1^h and J_2^h are marked by circular spots, crosses and open circles, respectively. The cubature formula can be written in the form

$$\int_Q u(x)dx = \sum_{k=0}^{m} \sum_{j\in J_k^h} \sum_{|q|=0}^{s} h^q u^{(q)}\left((j+\underline{1})h\right) \int_Q \tilde{\omega}_q\left(\frac{x}{h} - j\right)dx + \rho_h \,. \tag{6}$$

The formula is most easily calculated when $k = m$. Hence, it is sufficient to integrate over $\tilde{Q}_j$. Let $q = (q_1, q_2, \ldots, q_m)$, $j = (j_1, j_2, \ldots, j_m)$; then

$$\int_Q \tilde{\omega}_q\left(\frac{x}{h} - j\right)dx = \prod_{\nu=1}^{m} \int_0^1 \omega_{q_\nu}\left(\frac{x_\nu}{h} - j_\nu\right)dx_\nu = h^m \prod_{\nu=1}^{m} \int_0^2 \omega_{q_\nu}(t)dt \,.$$

If $q \neq 0$, then at least one of the $q_\nu \neq 0$ and the corresponding integral vanishes. If $q = 0$, then

$$\int_0^2 \omega_0(t)dt = 1 \,,$$

and finally

$$\sum_{j\in J_m^h} \sum_{|q|=0}^{s} h^q u^{(q)}\left((j+\underline{1})h\right) \int_Q \tilde{\omega}_q\left(\frac{x}{h} - j\right)dx = h^m \sum_{j\in J_m^h} u\left((j+\underline{1})h\right)$$

$$= h^m \sum_{j=0}^{\overline{2n-2}} u\left((j+\underline{1})h\right) = h^m \sum_{j=1}^{\overline{2n-1}} u(jh) \,. \tag{7}$$

We now consider those sums in (6) for which $k < m$. We take any vertix $j \in J_k^h$ and consider the corresponding larger cube Q_j : 2^k of its small cubes lie in Q . Obviously, each of these small cubes adjoins some k-dimensional edge of the cube Q and adjoins no edge of smaller dimension (see Figure 17). Let the equations of the mentioned k-dimensional edge be

$$x_{n_\mu} = \varepsilon_\mu , \quad \mu = 1, 2, \ldots, m-k ,$$

where ε_μ is either zero or unity. The other coordinates, denoted by $x_{p_1}, x_{p_2}, \ldots, x_{p_k}$, change on the intervals of length $2h$ in the domain $Q_j \cap Q$. More precisely, the coordinates x_{p_l} change on the interval $\left(j_{p_l} h, \left(j_{p_l}+2\right)h\right)$. If at least one of the components of the multi-index q , with value p_l , differs from zero, then

$$\int_Q \tilde{\omega}_q\left(\frac{x}{h} - j\right)dx = 0 ,$$

and the considered sum contains only items with $q_{p_1} = q_{p_2} = \ldots = q_{p_k} = 0$. It can be easily seen that

$$\int_Q \tilde{\omega}_q\left(\frac{x}{h} - j\right) = \begin{cases} 2^{-(m-k)}h^m , & q = \underline{0} , \\ h^m \prod_{\mu=1}^{m-k} (-1)^{1-\varepsilon_\mu} b_{n_\mu} , & q \neq 0 , \end{cases}$$

and the cubature formula finally takes the form

$$\int_Q u(x)dx = h^m \sum_{j\in J_m^h} u\big((j+1)h\big) + h^m \sum_{k=0}^{m-1} \sum_{j\in J_k^h} \frac{1}{2^{m-k}} u\big((j+1)h\big) +$$
$$+ h^m \sum_{k=0}^{m-1} \sum_{j\in J_k^h} \sum_{|q|=1}^{s} \prod_{\mu=1}^{m-k} (-1)^{1-\varepsilon_\mu} b_{n_\mu} h^q u^{(q)}\big((j+1)h\big) + \rho_h . \quad (8)$$

For a given $j \in J_k^h$, the summation in (8) is taken only over those q for which $q_{p_1} = q_{p_2} = \ldots = q_{p_k} = 0$.

§4. Integration Over a Ball

Let K be the unity ball in $\mathbb{R}_m$. Introducing the spherical coordinates $r, \vartheta_1, \vartheta_2, \ldots, \vartheta_{m-2}, \vartheta_{m-1}$, centered at the ball's centre, we map the ball onto the parallelpiped

$$0 < r < 1 \ , \quad 0 < \vartheta_k < \pi \ , \quad 1 \le k \le m-2 \ , \quad 0 < \vartheta_{m-1} < 2\pi \ .$$

Putting

$$\vartheta_k = \pi\Theta_k \ , \quad 1 \le k \le m-2 \ , \quad \vartheta_{m-1} = 2\pi\Theta_{m-1} \ ,$$

we map the ball K onto the unity cube Q defined by the inequalities

$$0 < r < 1 \ , \quad 0 < \Theta_k < 1 \ , \quad 1 \le k \le m-1 \ . \tag{1}$$

We locate the origin of the Cartesian coordinates at the centre of the ball K. Then the transformations are given by

$$\begin{aligned} x_1 &= r \cos \pi\Theta_1 \ , \\ x_2 &= r \sin \pi\Theta_1 \cos \pi\Theta_2 \ , \\ &\dots\dots\dots\dots \\ x_{m-1} &= r \sin \pi\Theta_1 \sin \pi\Theta_2 \dots \sin \pi\Theta_{m-2} \cos 2\pi\Theta_{m-1} \ , \\ x_m &= r \sin \pi\Theta_1 \sin \pi\Theta_2 \dots \sin \pi\Theta_{m-2} \sin 2\pi\Theta_{m-1} \ . \end{aligned} \tag{2}$$

The Jacobian of the transform (2) is

$$J = \frac{D(x_1, x_2, \dots, x_m)}{D(r, \Theta_1, \dots, \Theta_{m-1})} = 2\pi^{m-1} r^{m-1} \sin^{m-2} \pi\Theta_1 \sin^{m-3} \pi\Theta_2 \dots \sin \pi\Theta_{m-2} \ . \tag{3}$$

Let $u \in W_p^{(s+1)}(K)$. We evaluate the integral

$$L = \int_K u(x) dx \ . \tag{4}$$

Using the transform (2), it becomes

$$\int_Q V(r, \Theta_1, \Theta_2, \dots, \Theta_{m-1}) dr d\Theta_1 d\Theta_2 \dots d\Theta_{m-1} \ , \tag{5}$$

where $V = uJ$. It is not difficult to check that in the new variables $r, \Theta_1, \dots, \Theta_{m-1}$, $V \in W_p^{(s+1)}(K)$, and therefore the results of the preceding section can be applied to (5).

The same is true when the integration is taken not over the whole ball, but over any part bounded by surfaces of the form $r = \text{const.}$, $\Theta_1 = \text{const.}$, and so on.

We examine in details the case of the circle, which we denote by K. Let $u(x) = u(x_1, x_2) \in W_p^{(s+1)}(K)$. We evaluate

$$L = \int_K \int u(x_1, x_2) dx_1 dx_2 \ . \tag{6}$$

We put $x_1 = r \cos 2\pi\Theta$ and $x_2 = r \sin 2\pi\Theta$ and thus transform (6) into

$$L = 2\pi \int_0^1 \int_0^1 V(\Theta, r)\,dr\,d\Theta \qquad (7)$$

where $V(\Theta, r) = ru(r\cos 2\pi\Theta, r\sin 2\pi\Theta)$. We construct the square grid with grid size h on the plane (Θ, r) and apply the approximation (3.8) to the integral (7). For convenience, we introduce the notation $\Theta = y_1$, $r = y_2$, $j = (j_1, j_2)$. The first term in (3.8) takes the form

$$2\pi h^2 \sum_{j_1, j_2=0}^{2n-2} V\big((j+\underline{1})h\big) = 2\pi h^2 \sum_{j_1, j_2=0}^{2n-1} V\big(j_1 h, j_2 h\big) . \qquad (8)$$

This sum corresponds to gird points $j \in J_2^h$ marked by open circles on Figure 17.

Consider the set J_1^h (marked by crosses on Figure 17). We pick out the terms with $q = \underline{0}$. Since $b_0 = \frac{1}{2}$, these terms yield

$$2\pi \tfrac{1}{2} h^2 \sum_{j=1}^{2n-1} [V(jh, 0)+V(jh, 1)+V(0, jh)+V(1, jh)] .$$

At the centre of the circle, we have $V(\Theta, 0) = ru|_{r=0} = 0$. In addition, by periodicity, $V(0, r) = V(1, r)$, so the last sum can be simplified and rewritten as

$$2\pi h^2 \sum_{j=1}^{2n-1} [\tfrac{1}{2}V(jh, 1)+V(0, jh)] . \qquad (9)$$

The remaining terms of the sum, taken over J_1^h, yield

$$2\pi \sum_{j\in J_1^h} \sum_{|q|=0}^{s} (-1)^{1-\varepsilon_\mu} b_{n_\mu} h^q V^{(q)}\big((j+\underline{1})h\big) .$$

We have $y_1 = 0$ along the left hand side of the square, so $n_1 = 1$ and $\varepsilon_1 = 0$. We recall that in this situation we should evaluate the summation with $q_2 = 0$ and thereby obtain

$$-2\pi \sum_{j_2=0}^{2n-2} \sum_{q_1=1}^{s-1} b_{q_1} h^{q_1} V^{(q_1,0)} \big(0, (j_2+1)h\big) .$$

We obtain the same expression, but with a sign change, along the right hand side of the square using the fact that $\varepsilon_1 = 1$ and V is periodic. Thus, the sums along the vertical sides of the square vanish. The sums along the horizontal sides can be analogously reduced to the form

$$2\pi h^2 \sum_{q_2=1}^{s-1} \sum_{j_1=1}^{2n-1} b_{q_2} h^{q_2} \left[\frac{\partial^{q_2} V(j_1 h_1, 1)}{\partial \Theta^{q_2}} - \frac{\partial^{q_2} V(j_1 h, 0)}{\partial \Theta^{q_2}} \right] . \tag{10}$$

We note that the subtraction in (10) vanishes.

Finally, it remains to consider the sum over J_0^h (the four circled points on Figure 17). We have the following expression for $q = \underline{0}$

$$2\pi \frac{h^2}{4} [V(0, 0)+V(0, 1)+V(1, 0)+V(1, 1)] = \pi h^2 V(0, 1) . \tag{11}$$

The terms with $q \neq 0$ give

$$2\pi h^2 \sum_{|q|=1}^{s} b_{q_1} b_{q_2} h^q [V^{(q)}(0, 0)-V^{(q)}(0, 1)-V^{(q)}(1, 0)+V^{(q)}(1, 1)] ,$$

which vanishes by the periodicity of V. Summing (8)-(11), we obtain the cubature formula

$$\int_K \int u(x_1, x_2) dx_1 dx_2 = 2\pi h^2 \left[\sum_{j_1, j_2=1}^{2n-1} V(j_1 h, j_2 h) + \sum_{j=1}^{2n-1} V(jh, 1) + 2 \sum_{j=1}^{2n-1} V(0, jh) \right.$$
$$\left. + \tfrac{1}{2} V(0, 1) + \sum_{q=1}^{s-1} \sum_{j=1}^{2n-1} b_q h^q \frac{\partial^q V(jh, 1)}{\partial \Theta^q} \right] . \tag{12}$$

CHAPTER XI

ON INTEGRAL EQUATIONS

Approximate solutions of differential and integral equations of mathematical physics are most frequently constructed using direct methods, which enable the construction of the approximations to be reduced to the solution of linear algebraic system (assuming the given problem is linear). Direct methods are widely used. Different forms of the finite-difference method are the most common. A particular form of the variational-difference method was studied in Chapters VII-IX of this book. But, along with their merits, direct methods have one significant shortcoming: in order to improve the accuracy of an approximation, it is necessary to increase the order of the algebraic system, but this in turn leads to an increase in the instability of the associated numerical process. As we saw in Chapter IX, an increase in the order of the variational-difference algebraic system leads to a growth in its condition number, where the latter entails the accumulation of error during the solution of the algebraic system, and thereby implies a loss of accuracy. We examined this phenomenon in an example in §7 of Chapter IX. It is also important to remember that the computational solution of high order algebraic systems can involve storage problems.

The above considerations are the motivation for attempts to use methods, for the approximate solution of the equations of mathematical physics, which obviate the necessity to solve high order algebraic systems. They include iterative methods as well as methods based on recurrence relations. Below, we show how a comparatively simple example of such a method can be constructed for the approximate solutions of Fredholm integral equations of the second kind. A variational-difference approximation for the kernel is used. Initially, consider one-dimensional integral equations. Subsequently, we briefly discuss the extension of these results to some classes of multi-dimensional integral equations. The essence of our method is the construction of approximations for the resolvent of the given integral equation.

The basic result of the present chapter can be found in Mikhlin [19], [20].

Below, we use the same letter C to denote different constants. Their exact values are immaterial from the point of view of the present examination.

§1. Approximation of the Kernel and Resolvent

Consider the integral equation

$$u(x) - \lambda \int_0^1 K(x, y)u(y)dy = f(x) . \tag{1}$$

Its resolvent is defined by

$$\Gamma(x, y; \lambda) = \frac{D(x,y;\lambda)}{D(\lambda)} \tag{2}$$

where the "Fredholm determinant" $D(\lambda)$ and the "first Fredholm minor" $D(x, y; \lambda)$ are entire functions of λ. For the coefficients in their corresponding power series

$$D(\lambda) = \sum_{k=0}^{\infty} \frac{(-1)^k}{k!} c_k\lambda^k , \tag{3}$$

$$D(x, y; \lambda) = \sum_{k=0}^{\infty} \frac{(-1)^k}{k!} B_k(x, y)\lambda^k , \tag{4}$$

the following recurrence relations are known

$$c_0 = 1 , \quad c_k = \int_0^1 B_{k-1}(x, x)dx , \tag{5}$$

$$B_0(x, y) = K(x, y) , \quad B_k(x, y) = c_kK(x, y) - k \int_0^1 K(x, t)B_{k-1}(t, y)dt . \tag{6}$$

In addition, the coefficients of (4) are related to the kernel $K(x, y)$ by

$$B_k(x, y) = \int_0^1 \cdots \int_0^1 \begin{vmatrix} K(x, y) & K(x, t_1) & \cdots & K(x, t_k) \\ K(t_1, t) & K(t_1, t_1) & \cdots & K(t_1, t_k) \\ \cdots & \cdots & \cdots & \cdots \\ K(t_k, t_1) & K(t_k, t_2) & \cdots & K(t_k, t_k) \end{vmatrix} dt_1 \cdots dt_k . \tag{7}$$

The derivation of the relations (2)-(7), which were obtained by Fredholm himself, can be found in any "classic" text on integral equations. See, for example, Smirnov [1] or Lovitt [1].

The solutions of (1) can be approximated in the following way. We approximate $K(x, y)$ by the following degenerate kernel

$$K(x, y) \approx K'(x, y) = \sum_{j=1}^{n} a_j(x) b_j(y) \tag{8}$$

and construct the resolvent of $K'(x, y)$ by calculating the corresponding coefficients c_k' and $B_k'(x, y)$ using (5) and (6). In particular, we obtain

$$B_k'(x, y) = \sum_{i,j=1}^{n} \alpha_{ij}^{(k)} a_i(x) b_j(y)$$

and hence,

$$c_k' = \sum_{i,j=1}^{n} \alpha_{ij}^{(k)} \int_0^1 a_i(x) b_j(x) dx \tag{9}$$

where the coefficients $\alpha_{i,j}^{(k)}$ are defined by the recurrence relations

$$\alpha_{ij}^{(k)} = c_k' \delta_{ij} - k \sum_{l,m=1}^{n} \delta_{im} \alpha_{lj}^{(k-1)} \int_0^1 a_l(t) b_m(t) dt \ . \tag{10}$$

In this approach, the approximation of the resolvent involves the calculation of the n^2 integrals contained in (9) and (10). As n increases, the number of functions $a_i(x)$ and $b_i(y)$ increases, and therefore, it will be necessary to recalculate the mentioned integrals. This shortcoming can be eliminated through the use of a variational-difference approximation for the kernel.

We assume that the kernel $K \in C^{(2s)}(Q)$, where Q is the square $\{(x, y),\ 0 \le x,\ y \le 1\}$ and s is a positive integer. Where necessary, we extend K to $C^{(2s)}(\Omega)$, where Ω is some region containing Q. We introduce the one-dimensional primitive functions $\omega_0(x), \omega_1(x), \ldots, \omega_{s-1}(x)$ described in §6 of Chapter III, which give the highest order of approximation, and construct the approximate kernel

$$K^h(x, y) = \sum_{\underline{q}=\underline{0}}^{\underline{s-1}} \sum_{\underline{j}=\underline{-1}}^{\underline{2n-1}} b_{qjh}^{(0)} \tilde{\omega}_q\left(\frac{z}{h} - j\right) \tag{11}$$

where $h = 1/2n$ with n a positive integer. In (11), q and j denote two-dimensional multiindices (namely, $q = (q_1, q_2)$, $j = (j_1, j_2)$), and z is a general planar point with coordinates x, y. In addition,

$$\tilde{\omega}_q(z) = \omega_{q_1}(x) \omega_{q_2}(y) \ , \tag{12}$$

$$b_{qjh}^{(0)} = \begin{cases} h^q K^{(q)}\left((j_1+1)h,\ (j_2+1)h\right) \ , & |q| \le s \ , \\ 0 \ , & |q| \ge s \ . \end{cases} \tag{13}$$

We introduce the notation $b_{qjh}^{(0)} = b_{q_1q_2j_1j_2h}^{(0)}$. As we proved in §6 of Chapter III,

$$|K^h(x, y)-K(x, y)| \le Ch^{2s} . \tag{14}$$

We now derive for $K^h(x, y)$ formulas for the coefficients c_k and $B_k(x, y)$. We denote them by c_k^h and $B_k^h(x, y)$, respectively. We assume that, for some k,

$$B_{k-1}^h(x, y) = \sum_{\underline{q}=\underline{0}}^{\underline{s}-\underline{1}} \sum_{\underline{j}=-\underline{1}}^{\underline{2n}-\underline{1}} b_{qjh}^{(k-1)}\tilde{\omega}_q\left(\frac{z}{h} - j\right) . \tag{15}$$

Then, obviously,

$$c_k^h = \sum_{\underline{q}=\underline{0}}^{\underline{s}-\underline{1}} \sum_{\underline{j}=-\underline{1}}^{\underline{2n}-\underline{1}} b_{qjh}^{(k-1)} \int_0^1 \omega_{q_1}\left(\frac{t}{h} - j_1\right)\omega_{q_2}\left(\frac{t}{h} - j_2\right)dt \tag{16}$$

and

$$B_k^h(x, y) = \sum_{\underline{q}=\underline{0}}^{\underline{s}-\underline{1}} \sum_{\underline{j}=-\underline{1}}^{\underline{2n}-\underline{1}} b_{qjh}^{(k)}\omega_q\left(\frac{z}{h} - j\right) \tag{17}$$

where the $b_{qjh}^{(k)}$ denote the new coefficients. We introduce the notation

$$\int_0^1 \omega_{q_1}\left(\frac{t}{h} - j_1\right)\omega_{q_2}\left(\frac{t}{h} - j_2\right)dt = hw_{qj} = hw_{q_1q_2j_1j_2} . \tag{18}$$

Using (6), (11) and (15), we find that

$$B_k^h(x, y) = \sum_{\underline{q}=\underline{0}}^{\underline{s}-\underline{1}} \sum_{\underline{j}=-\underline{1}}^{\underline{2n}-\underline{1}} c_k^h b_{qjh}^{(0)}\omega_{q_1}\left(\frac{x}{h} - j_1\right)\omega_{q_2}\left(\frac{y}{h} - j_2\right)$$
$$- k \sum_{\underline{q},\underline{r}=\underline{0}}^{\underline{s}-\underline{1}} \sum_{\underline{j},\underline{l}=-\underline{1}}^{\underline{2n}-\underline{1}} \int_0^1 b_{qjh}^{(0)}b_{rlh}^{(k-1)}\omega_{q_1}\left(\frac{x}{h} - j_1\right)\omega_{r_2}\left(\frac{y}{h} - l_2\right)\omega_{q_2}\left(\frac{t}{h} - j_2\right)\omega_{r_1}\left(\frac{t}{h} - l_1\right)dt .$$

We interchange r_2 and q_2, as well as j_2 and l_2, under the integral sign. Using (18) we find

$$b_{qjh}^{(k)} = c_k^h b_{qjh}^{(0)} - kh \sum_{\underline{r}=\underline{0}}^{\underline{s}-\underline{1}} \sum_{\underline{l}=-\underline{1}}^{\underline{2n}-\underline{1}} b_{q_1r_2j_1l_2h}^{(0)} b_{r_1q_2l_1j_2h} w_{rl} . \tag{19}$$

We examine the integral (18) in some detail. It vanishes, if the supports of the terms under the integral sign do not intersect, and this occurs when $|j_1-j_2| > 1$. It is sufficient to consider the integrals with $j_1 - j_2 = -1, 0, 1$; that is, the integrals

$$\int_0^1 \omega_{q_1}\left(\frac{t}{h} - m\right)\omega_{q_2}\left(\frac{t}{h} - m\right)dt = h\omega_{q_1q_2mm}\,,$$
$$\int_0^1 \omega_{q_1}\left(\frac{t}{h} - m\right)\omega_{q_2}\left(\frac{t}{h} - m + 1\right)dt = h\omega_{q_1q_2m,m-1}\,. \tag{20}$$

For different values of m we obtain the following values for $\omega_{q_1q_2mm}$:

$$m = -1 : \quad \omega_{q_1q_2,-1,-1} = \int_0^1 \omega_{q_1}(t+1)\omega_{q_2}(t+1)dt \tag{20_1}$$

$$m = 2n - 1 : \quad \omega_{q_1q_2,2n-1,2n-1} = \int_0^1 \omega_{q_1}(t)\omega_{q_2}(t)dt \tag{20_2}$$

$$0 \le m \le 2n-2 : \quad \omega_{q_1q_2mm} = \int_0^2 \omega_{q_1}(t)\omega_{q_2}(t)dt\,. \tag{20_3}$$

We note that the integral (20_3) is the sum of integrals (20_1) and (20_2). In addition,

$$\omega_{q_1q_2m,m-1} = \int_0^1 \omega_{q_1}(t)\omega_{q_2}(t+1)dt\,. \tag{20_4}$$

Thus, for the calculation of the coefficients in the Fredholm series, it is sufficient to evaluate the integrals (20_1), (20_2) and (20_4). For a given s the total number of such integrals equals $s(2s+1)$. In addition, they depend neither on h nor on the kernel K. This is the advantage of the variational-difference approximation when compared with the general approximations considered at the beginning of this section. We note that the integrals (20) are basically elementary, since the primitive functions are polynomials of degree $2s - 1$ on each of the intervals $(0, 1)$ and $(1, 2)$. The mentioned integrals can be calculated in advance, and then stored for subsequent use.

§2. The Accuracy of the Approximation

We estimate the error in the Fredholm series, arising from the replacement of $K(x, y)$ by $K^h(x, y)$. We assume that $|K(x, y)| < 1$ - this can always be achieved by changing the parameter λ in (1.1). Then, for sufficiently small h, we have that $|K^h(x, y)| < 1$. We denote by Δ_{k+1} the determinant under the integral sign in (1.7), and by Δ_{k+1}^h the analogous determinant for the kernel $K^h(x, y)$. The difference $\Delta_{k+1} - \Delta_{k+1}^h$ can be expressed as the following sum of $k + 1$ determinants

$$\Delta_{k+1} - \Delta_{k+1}^{h} = \begin{vmatrix} K(x, y)-K^{h}(x, y) & K(x, t_1) & \dots & K(x, t_k) \\ K(t_1, y)-K^{h}(t_1, y) & K(t_1, t_1) & \dots & K(t_1, t_k) \\ \dots & \dots & \dots & \dots \\ K(t_k, y)-K^{h}(t_k, y) & K(t_k, t_1) & \dots & K(t_k, t_k) \end{vmatrix}$$

$$+ \sum_{\nu=1}^{k} \begin{vmatrix} K^{h}(x, y) & \dots & K^{h}(x, t_{\nu-1}) & K(x, t_\nu)-K^{h}(x, t_\nu) & K(x, t_{\nu+1}) & \dots & K(x, t_k) \\ K^{h}(t_1, y) & \dots & K^{h}(t_1, t_{\nu-1}) & K(t_1, t_\nu)-K^{h}(t_1, t_\nu) & K(t_1, t_{\nu+1}) & \dots & K(t_1, t_k) \\ \dots & \dots & \dots & \dots & \dots & \dots & \dots \\ K^{h}(t_k, y) & \dots & K^{h}(t_k, t_{\nu-1}) & K(t_k, t_\nu)-K^{h}(t_k, t_\nu) & K(t_k, t_{\nu+1}) & \dots & K(t_k, t_k) \end{vmatrix} . \tag{1}$$

For the determinant

$$\Delta = \begin{vmatrix} a_{11} & a_{12} & \dots & a_{1,k+1} \\ a_{21} & a_{22} & \dots & a_{2,k+1} \\ \dots & \dots & \dots & \dots \\ a_{k+1,1} & a_{k+1,2} & \dots & a_{k+1,k+1} \end{vmatrix}$$

we have, using Hadamard Theorem,

$$|\Delta| \le \sqrt{\prod_{i=1}^{k+1} \sum_{j=1}^{k+1} a_{ij}^2} .$$

In the present context, the sum is dominated by $k + 1$. But, inequality (1.14) shows that the individual terms are dominated by $C(k+1)h^{4s}$. Thus, each determinant in (1) is dominated by $Ch^{2s}(k+1)^{(k+1)/2}$. Hence,

$$\left|\Delta_{k+1}-\Delta_{k+1}^{h}\right| \le Ch^{2s}(k+1)^{(k+3)/2} .$$

After integration, we find that

$$\left|B_k(x, y)-B_k^{h}(x, y)\right| \le Ch^{2s}(k+1)^{(k+3)/2} \tag{2}$$

which indicates that the errors in both Fredholm series are dominated by the same series

$$Ch^{2s} \sum_{k=0}^{\infty} \frac{|\lambda|^{k}(k+1)^{(k+3)/2}}{k!} . \tag{3}$$

We estimate the sum of the series (3). By Stirling's formula

$$k! = \sqrt{2\pi k} \left(\frac{k}{e}\right)^{k} e^{\Theta/12k} , \quad 0 < \Theta < 1 .$$

Hence, $k^{(k+1)/2} \le (2\pi)^{-\frac{1}{2}} k! e^{k}$ and, consequently

$$(k+1)^{(k+3)/2} < C(k+1)^{5/4}\sqrt{k!}\, e^{k/2} .$$

Thus, the sum of the series (3) is dominated by

$$\sum_{k=0}^{\infty} \frac{(k+1)^{5/4}\sigma^{k/2}}{\sqrt{k!}} \le \left[\sum_{k=0}^{\infty} \frac{1}{(k+1)^{3/2}}\right]^{\frac12} \left[\sum_{k=0}^{\infty} \frac{(k+1)^4\sigma^k}{k!}\right]^{\frac12} = C\left[\sum_{k=0}^{\infty} \frac{(k+1)^4\sigma^k}{k!}\right]^{\frac12} , \quad \sigma = e|\lambda|^2 .$$

Picking out the terms with $k = 0, 1, 2, 3$, we obtain (letting P_3 denote a polynomial of degree 3)

$$\sum_{k=0}^{\infty} \frac{(k+1)^4\sigma^k}{k!} = P_3\left(|\lambda|^2\right) + \sigma^4 \sum_{k=4}^{\infty} \frac{(k+1)^4\sigma^{k-4}}{k!} \le P_3\left(|\lambda|^2\right) + C|\lambda|^8 \sum_{k=4}^{\infty} \frac{k(k-1)\dots(k-3)\sigma^{k-4}}{k!}$$

$$= P_3\left(|\lambda|^2\right) + C|\lambda|^8 \exp\left(e|\lambda|^2\right) .$$

Finally,

$$\begin{aligned} |D(x, y; \lambda)-D^h(x, y; \lambda)| &\le Ch^{2s}\left[P_3\left(|\lambda|^2\right)+|\lambda|^8\exp\left(e|\lambda|^2\right)\right]^{\frac12} , \\ |D(\lambda)-D^h(\lambda)| &\le Ch^{2s}\left[P_3\left(|\lambda|^2\right)+|\lambda|^8\exp\left(e|\lambda|^2\right)\right]^{\frac12} , \end{aligned} \qquad (4)$$

where $D^h(\lambda)$ and $D^h(x, y; \lambda)$ denote the Fredholm determinant and the first Fredholm minor for the kernel $K^h(x, y)$.

We derive an additional estimate. Truncate the series for $D^h(\lambda)$ and $D^h(x, y; \lambda)$ after terms involving powers of λ less than or equal to N . Then, the remainder associated with the truncation of $D^h(x, y; \lambda)$ has the form

$$\sum_{k=N+1}^{\infty} \frac{|\lambda|^k \cdot |B_k^h(x,y)|}{k!} \le \frac{C}{\sqrt[4]{N-1}} \cdot \frac{|\lambda|^{N+2} e^{N/2+1}}{\sqrt{(N-1)!}} \exp\left(\tfrac12 e|\lambda|^2\right) . \qquad (5)$$

An analogous estimate holds for the remainder associated with the truncation of $D^h(\lambda)$.

§3. Rounding Error Accumulation - Absolute Estimates

In the next two sections, we analyse the rounding error accumulation occuring in the evaluation of the coefficients $b_{qjh}^{(0)}$ in (1.11), $b_{qjh}^{(k)}$ in (1.19) and C_k^h in (1.16).

Let $\delta_{qjh}^{(0)}$ denote the accumulated error arising in the evaluation of the coefficient $b_{qjh}^{(0)}$; namely, $b_{qjh}^{(0)} = \bar{b}_{qjh}^{(0)} + \delta_{qjh}^{(0)}$, where $\bar{b}_{qjh}^{(0)}$ denotes the known nonexact value of $b_{qjh}^{(0)}$. Obviously, we assume that $\delta_{qjh}^{(0)} = 0$ for $|q| > s$. For such values of q we also have $\bar{b}_{qjh}^{(0)} = 0$. We write

$$\bar{K}^h(x, y) = \sum_{q=\underline{0}}^{\underline{s-1}} \sum_{j=\underline{-1}}^{\underline{2n-1}} \bar{b}_{qjh}^{(0)} \tilde{\omega}\left(\frac{z}{h} - j\right) . \tag{1}$$

The kernel $\bar{K}^h(x, y)$ is known exactly. It follows from (1.14) that

$$|\bar{K}^h(x, y) - K(x, y)| \le C\left(h^{2s} + \delta_0\right) \tag{2}$$

where δ_0 is the supremum of the values $\left|\delta_{qjh}^{(0)}\right|$. The value δ_0 is known, since, from a practical point of view, the errors $\delta_{qjh}^{(0)}$ arise as a result of approximating the coefficients $b_{qjh}^{(0)}$ by finite decimal, binary or some other rational fraction representation. Below, we denote by $\bar{K}^h(x, y)$ the given variational-difference approximation of the kernel $K(x, y)$: assume that the coefficients $b_{qjh}^{(0)}$ are known exactly (without error). The resolvent of this new kernel is

$$\bar{\Gamma}^h(x, y) = \frac{\bar{D}^h(x, y; \lambda)}{\bar{D}^h(\lambda)}$$

where

$$\bar{D}^h(x, y; \lambda) = \sum_{k=0}^{\infty} \frac{(-1)^k \bar{B}_k^h(x, y)}{k!} \lambda^k , \quad \bar{D}^h(\lambda) = \sum_{k=0}^{\infty} \frac{(-1)^k \bar{C}_k^h}{k!} \lambda^k ,$$

with

$$\bar{B}_k^h(x, y) = \bar{C}_k^h \bar{K}^h(x, y) - k \int_0^1 \bar{K}^h(x, t) \bar{B}_{k-1}^h(t, y) dt ,$$

$$\bar{C}_k^h = \int_0^1 \bar{B}_{k-1}^h(t, t) dt .$$

The last can be expressed in terms of primitive functions (see §1)

$$\bar{B}_k^h(x, y) = \sum_{q=\underline{0}}^{\underline{s-1}} \sum_{j=\underline{-1}}^{\underline{2n-1}} \bar{b}_{qjh}^{(k)} \tilde{\omega}_q\left(\frac{z}{h} - j\right) ,$$

$$\bar{C}_k^h = h \sum_{q=\underline{0}}^{\underline{s-1}} \sum_{j=\underline{-1}}^{\underline{2n-1}} \bar{b}_{qjh}^{(k-1)} \omega_{qj} , \tag{3}$$

$$\bar{b}^{(k)}_{qjh} = \bar{c}^{h}_{k}\bar{b}^{(0)}_{qjh} - kh \sum_{r=0}^{s-1} \sum_{l=-1}^{2n-1} \bar{b}^{(0)}_{q_1r_2j_1l_2h}\bar{b}^{(k-1)}_{r_1q_2l_1j_2h}w_{lr} \; . \tag{4}$$

As we saw above $w_{lr} = 0$, when $|l_1 - l_2| > 1$. Hence, the number of non-zero terms in (3) and (4) has order $O(n) = O(h^{-1})$. Consequently, the coefficients $\bar{b}^{(k)}_{qjh}$ and $\bar{c}^{h}_{k}$ are bounded independently of h (obviously, the last holds under the assumption that these coefficients are evaluated exactly using (3) and (4)). Let M be any number larger than $\|K\|_{C^{(s-1)}(Q)}$. We choose h and δ_0 sufficiently small so that $\|\bar{K}^{h}\|_{C^{(s-1)}(Q)} \le M$. We set

$$\frac{\bar{b}^{(k)}_{qjh}}{k!} = \beta^{(k)}_{qjh} \, , \quad \frac{\bar{c}^{h}_{k}}{k!} = \gamma^{h}_{k} \; . \tag{5}$$

The recurrence relations for these new values are

$$\gamma^{h}_{k} = \frac{h}{k} \sum_{q=0}^{s-1} \sum_{j=-1}^{2n-1} \beta^{(k-1)}_{qjh} w_{qj} \; , \tag{6}$$

$$\beta^{(k)}_{qjh} = \gamma^{h}_{k}\beta^{(0)}_{qjh} - h \sum_{r=0}^{s-1} \sum_{l=-1}^{2n-1} \beta^{(0)}_{q_1r_2j_1l_2h}\beta^{(k-1)}_{r_1q_2l_1j_2h}w_{rl} \; . \tag{7}$$

We denote the errors associated with $\beta^{(k)}_{qjh}$ and γ^{h}_{k} by $\delta^{(k)}_{qjh}$ and $\tilde{\delta}^{h}_{k}$, respectively. For simplicity of presentation, we assume that the w_{qj} are evaluated exactly, so that there are only two sources of errors: the errors $\delta^{(k-1)}_{qjh}$ in $\beta^{(k-1)}_{qjh}$ and the rounding error accumulation in the sums in (8) and (9). We denote by $\tilde{\varepsilon}^{h}_{k}$ the accumulated error in the second term on the r.h.s. of (8), and by $\varepsilon^{(k)}_{qjh}$ the accumulated error in the second term on the r.h.s. of (9). As a consequence we obtain

$$\tilde{\delta}^{h}_{k} = \frac{h}{k} \sum_{q=0}^{s-1} \sum_{j=-1}^{2n-1} \delta^{(k-1)}_{qjh} w_{qj} + \tilde{\varepsilon}^{h}_{k} \; , \tag{8}$$

$$\delta^{(k)}_{qjh} = \tilde{\delta}^{h}_{k}\beta^{(0)}_{qjh} - h \sum_{r=0}^{s-1} \sum_{l=-1}^{2n-1} \beta^{(0)}_{q_1r_2j_1l_2h}\delta^{(k-1)}_{r_1q_2l_1j_2h}w_{rl} + \varepsilon^{(k)}_{qjh} \; . \tag{9}$$

Let $\delta_{kh} = \max_{q,j} \left|\delta^{(k)}_{qjh}\right|$, $\varepsilon_{kh} = \max_{q,j} \left[\left|\varepsilon^{(k)}_{qjh}\right| + \left|\tilde{\varepsilon}^{h}_{k}\beta^{(0)}_{qjh}\right|\right]$. The number of non-zero values w_{qj} has order $O(h^{-1})$, and the values of w_{qj} are bounded. Taking the latter into account, we obtain from (10) and (11) the new inequality (omitting the subscript h)

$$\delta_k \le c_0 M\left(1+\frac{1}{k}\right)\delta_{k-1} + \varepsilon_k \;, \quad c_0 = \text{const.} \tag{10}$$

The loss of accuracy associated with the accumulation of rounding errors depends to some extent on the computer being used. We therefore consider two appropriate choices for ε_k .

1. $\varepsilon_k = \varepsilon = \text{const.}$ Then

$$\delta_k \le c_0 M\left(1+\frac{1}{k}\right)\delta_{k-1} + \varepsilon \;.$$

Hence

$$\delta_k \le (c_0 M)^k \prod_{j=1}^{k}\left(1+\frac{1}{j}\right)\delta_0 + \varepsilon\Big[1 + c_0 M\left(1+\frac{1}{k}\right) + (c_0 M)^2\left(1+\frac{1}{k}\right) \times$$
$$\times \left(1+\frac{1}{k-1}\right) + \ldots + (c_0 M)^{k-1}\left(1+\frac{1}{k}\right)\left(1+\frac{1}{k-1}\right)\ldots\left(1+\frac{1}{2}\right)\Big] \;.$$

Assume that $k \ge 2$. We replace every term in round brackets by $3/2$. Then the mentioned coefficient is dominated by

$$1 + \frac{3c_0 M}{2} + \left(\frac{3c_0 M}{2}\right)^2 + \ldots + \left(\frac{3c_0 M}{2}\right)^{k-1} = \left(\left(\frac{3c_0 M}{2}\right)^k - 1\right) \Big/ \left(\frac{3c_0 M}{2} - 1\right)$$

and, consequently,

$$\delta_k \le \delta_0 (c_0 M)^k \prod_{j=1}^{k}\left(1+\frac{1}{j}\right) + \varepsilon\left(\left(\frac{3c_0 M}{2}\right)^k - 1\right) \Big/ \left(\frac{3c_0 M}{2} - 1\right) .$$

Now,

$$\ln \prod_{j=1}^{k}\left(1+\frac{1}{j}\right) = \sum_{j=1}^{k} \ln\left(1+\frac{1}{j}\right) < \sum_{j=1}^{k}\frac{1}{j} < \gamma + \ln(k+1) \;,$$

where γ denotes Euler's constant, so that $\prod_{j=1}^{k}\left(1+\frac{1}{j}\right) < (k+1)e^{\gamma}$ and

$$\delta_k < e^{\gamma}(c_0 M)^k (k+1)\delta_0 + \varepsilon\left(\left(\frac{3c_0 M}{2}\right)^k - 1\right) \Big/ \left(\frac{3c_0 M}{2} - 1\right) . \tag{11}$$

2. Let $\varepsilon_k = a\left(1+\frac{1}{k}\right)\delta_{k-1}$, $a = \text{const} > 0$. Then $\delta_k < b\left(1+\frac{1}{k}\right)\delta_{k-1}$, $b = a + c_0 M$, and hence,

$$\delta_k \le \delta_0 b^k \prod_{j=1}^{k}\left(1+\frac{1}{j}\right) < \delta_0 e^{\gamma}(k+1)b^k \;. \tag{12}$$

In this situation, the errors decrease more slowly than the geometric progression with multiplier b , if $b < 1$, and increase more rapidly than the same progression, if

$b > 1$.

§4. Rounding Error Accumulation - Probabilistic Estimates

The rather conservative estimates of §3 are only likely to be attained in pathological situations such as when all the errors $\delta_{qjh}^{(k)}$ are of the same sign. We therefore consider another possibility.

We allow for the possibility that partial cancellation of the errors in (3.10) and (3.11) occurs. Retaining the notations of §3, we formulate this as follows: the accumulated errors in the mentioned sums are dominated by $c_0 h^{\alpha-1}\delta_{k-1}$ and $c_0 M h^{\alpha-1}\delta_{k-1}$, respectively, where $\alpha = \text{const} > 0$. We then obtain the following inequalities from (3.10) and (3.11):

$$|\tilde{\delta}_k| \le \frac{c_0 M h^\alpha}{k}\,\delta_{k-1} + \tilde{\varepsilon}_k\ , \quad \delta_k \le c_0 M h^\alpha\left(1 + \frac{1}{k}\right)\delta^{k-1} + \varepsilon_k\ . \tag{1}$$

The second inequality in (1) differs from (3.12) only in that the multiplier $c_0 M$ is replaced by $c_0 M h^\alpha$. It can be easily proved that, for sufficiently small h , for example, for $h < (3c_0 M)^{-1/\alpha}$ and for $\varepsilon_k = \varepsilon = \text{const.}$, we have

$$\delta_k = \left(\frac{h^\alpha}{2}\right)^k (k+1)e^\gamma\delta_0 + 2\varepsilon < 3^{-k}(k+1)e^\gamma\delta_0 + 2\varepsilon\ , \tag{2}$$

and the accumulated error is bounded. Further, if

$$\varepsilon_k < ah^\alpha\left(1 + \frac{1}{k}\right)\delta_{k-1}\ , \quad a = \text{const.}, \tag{3}$$

then

$$\delta_k < (bh^\alpha)^k (k+1)e^\gamma\delta_0\ , \quad b = c_0 M + a\ , \tag{4}$$

and the error decreases sufficiently rapidly as k increases.

We estimate the probability that the inequalities for (3.10) and (3.11) are valid. We assume that the $\delta_{qj}^{(k)}$ are independent and identically distributed random variables with zero mean. We also assume that these variables are bounded $\left|\delta_{qj}^{(k)}\right| \le \delta_k$. Then their dispersions are bounded; namely, $\mathrm{D}\left(\delta_{qj}^{(k)}\right) \le \delta_k^2$. We denote the sums in (3.10) and (3.11) by $\tilde{\Delta}_q^k$ and $\tilde{\Delta}_{qj}^k$, respectively. We estimate the dispersions of these sums in the following way (c_1 denotes some constant)

$$\mathsf{D}\left(\tilde{\Delta}^{k}_{q}\right) = \frac{h^2}{k^2} \sum_{q=0}^{s-1} \sum_{j=-1}^{2n-1} w^2_{qj} \mathsf{D}\left(\delta^2_{qj}\right) \le \frac{c_1 h}{k^2} \delta^2_{k-1} , \tag{5}$$

$$\mathsf{D}\left(\tilde{\Delta}^{k}_{qj}\right) = h^2 \sum_{r=0}^{s-1} \sum_{l=-1}^{2n-1} \left(\beta^{(0)}_{q_1 r_2 j_1 l_2}\right)^2 w^2_{rl} \mathsf{D}\left(\delta^{(k-1)}_{r_1 q_2 l_1 j_2}\right) \le c_1^2 M^2 h \delta^2_{k-1} . \tag{6}$$

We make use of the Chebyschev inequality

$$\mathsf{P}\left(|\tilde{\Delta}^{k}_{qj}| \le t c_1 M\sqrt{h}\, \delta_{k-1}\right) \le 1 - t^{-2} .$$

We put $t = c_0 c_1^{-1} h^{\alpha-3/2}$ and obtain

$$\mathsf{P}\left(|\tilde{\Delta}^{k}_{qj}| \le c_0 M h^{\alpha-1} \delta_{k-1}\right) \le 1 - \bar{c}_0^{2} c_1^2 h^{3-2\alpha} . \tag{7}$$

In the same way, applying the Chebyschev inequality to $|\tilde{\Delta}^{k}_{q}|$ and putting $t = c_0 c_1^{-1} k h^{\alpha-3/2}$, we obtain

$$\mathsf{P}\left(|\tilde{\Delta}^{k}_{q}| \le c_0 h^{\alpha-1} \delta_{k-1}\right) \le 1 - \frac{c_1^2}{c_0^2 k^2} h^{3-2\alpha} . \tag{8}$$

If h is sufficiently small and $0 < \alpha < 3/2$, then the probabilities (7) and (8) are close to unity and, consequently, the hypothesis, formulated at the beginning of the present section is "a practical reality".

Now, if ε_k is chosen to be consistent with (3) and the parameter λ in the integral equation is such that $|\lambda| < (bh^{\alpha})^{-1}$, then the accumulated errors in the Fredholm series $\bar{D}^{h}(x, y; \lambda)$ and $\bar{D}^{h}(\lambda)$ are bounded with probability given by (7).

§5. Integral Equations Which Can Be Solved by Iteration

As before, let the kernel of the integral equation (1.1) be sufficiently smooth and of small absolute magnitude, so that the resolvent can be expressed as a series of iterated kernels

$$\Gamma(x, y; \lambda) = \sum_{k=0}^{\infty} \lambda^k K_{k+1}(x, y) . \tag{1}$$

We construct the kernel $\bar{K}^{h}(x, y)$ using (3.1) and choose the values h and δ_0 sufficiently small so that the series for the resolvent $\bar{\Gamma}^{h}(x, y; \lambda)$ of the kernel $\bar{K}^{h}(x, y)$ corresponding to (1) converges, and so that the difference between the resolvents $\Gamma(x, y; \lambda)$ and $\bar{\Gamma}^{h}(x, y; \lambda)$ is sufficiently small. Then the problem reduces to the evaluation of the iterated kernels $K_k^h(x, y)$.

It is convenient to change the notations for the coefficients in (3.1) by writing

$$\overline{K}^h(x, y) = \sum_{\underline{q}=\underline{0}}^{\underline{s}-\underline{1}} \sum_{\underline{j}=-\underline{1}}^{2\underline{n}-\underline{1}} a_{qj}^{(0)} \tilde{\omega}_q\left(\frac{z}{h} - j\right) , \quad a_{qj}^{(0)} = 0 , \quad |q| > s . \tag{2}$$

Obviously,

$$\overline{K}_k^h(x, y) = \sum_{\underline{q}=\underline{0}}^{\underline{s}-\underline{1}} \sum_{j=-1}^{2\underline{n}-\underline{1}} a_{qj}^{(k)} \tilde{\omega}_q\left(\frac{z}{h} - j\right) . \tag{3}$$

The formula

$$\overline{K}_k^h(x, y) = \int_0^1 \overline{K}^h(x, t)\overline{K}_{k-1}^h(t, y)dt$$

gives the recurrence relation for the coefficients $a_{qj}^{(k)}$ (compare with §1)

$$a_{qj}^{(k)} = h \sum_{\underline{r}=\underline{0}}^{\underline{s}-\underline{1}} \sum_{\underline{l}=-\underline{1}}^{2\underline{n}-\underline{1}} a_{q_1 r_2 j_1 l_2}^{(0)} a_{r_1 q_2 l_1 j_2}^{(k-1)} w_{rl} . \tag{4}$$

We study the error associated with (4). As in §4, we assume that the values $a_{qj}^{(0)}$ and w_{qj} are known exactly. Then the error in $a_{qj}^{(1)}$ arises from rounding, while the error in $a_{qj}^{(k)}$, $k > 1$, arises from rounding and from the error in $a_{qj}^{(k-1)}$. We denote by $\mu_{qj}^{(k)}$ the error in $a_{qj}^{(k)}$ and write $\mu_k = \max_{q,j} |\mu_{qj}^{(k)}|$. In addition, we denote by $\nu_{qj}^{(k)}$ the rounding error in (4) and write $\nu_k = \max_{q,j} |\nu_{qj}^{(k)}|$. From (4) it follows

$$\mu_{qj}^{(k)} = h \sum_{\underline{r}=\underline{0}}^{\underline{s}-\underline{1}} \sum_{\underline{l}=-\underline{1}}^{2\underline{n}-\underline{1}} a_{q_1 r_2 j_1 l_2}^{(0)} \mu_{r_1 q_2 l_1 j_2}^{(k-1)} w_{rl} + \nu_{qj}^{(k)} . \tag{5}$$

Hence

$$\mu_k \le c_0 M \mu_{k-1} + \nu_k , \quad c_0 = \text{const.} \tag{6}$$

As in §§3 and 4, we consider the following two situations:

1. $\nu_k \le \nu_0 = \text{const.}$ Then $\mu_k \le \tilde{\mu}_k$ where $\tilde{\mu}_k$ satisfies the equation $\tilde{\mu}_k = c_0 M \tilde{\mu}_{k-1} + \nu_0$, $\tilde{\mu}_1 = \mu_1$. The solution of this equation is

$$\tilde{\mu}_k = (c_0 M)^{k-1}\tilde{\mu}_1 + \nu_0 \frac{(c_0 M)^{k-1}-1}{c_0 M-1}$$

and, consequently,

$$\mu_k \le (c_0 M)^{k-1}\mu_1 + \nu_0 \frac{(c_0 M)^{k-1}-1}{c_0 M-1} . \tag{7}$$

Thus, the error increases like a geometric progression, if $c_0 M > 1$, and is bounded, if $c_0 M < 1$. When $c_0 M = 1$, the error estimate becomes $\mu_k \le \mu_1 + (k-1)\nu_0$.

2. $\nu_k \le c_0 M\mu_{k-1}$. In this situation, $\mu_k \le 2c_0 M\mu_{k-1}$ and

$$\mu_k \le (2c_0 M)^{k-1}\mu_1 . \tag{8}$$

Thus, the error increases like a geometric progression, if $2c_0 M > 1$, is bounded, if $2c_0 M = 1$, and decreases like a geometric progression, if $2c_0 M < 1$.

The estimate improves, if we assume that partial cancellation of the errors in (5) occurs. The estimate reduces to

$$\left|\sum_{r=0}^{s-1}\sum_{l=-1}^{2n-1} a^{(0)}_{q_1 r_2 j_1 l_2}\mu_{r_1 q_2 l_1 j_2}\omega_{rl}\right| \le c_0 Mh^{\alpha-1}\mu_{k-1} , \quad 0 < \alpha < 1 . \tag{9}$$

Then $\mu_k \le c_0 Mh^{\alpha}\mu_{k-1} + \nu_k$. For sufficiently small h , we have $c_0 Mh^{\alpha} < 1$. By (7),

$$\mu_k \le \left(c_0 Mh^{\alpha}\right)\mu_1 + \nu_0 \frac{\left(c_0 Mh^{\alpha}\right)^{k-1}-1}{\hat{c}_0 Mh^{\alpha}-1}$$

and, consequently, for $\nu_k \le \nu_0 = \text{const}$, the errors are bounded. If we assume $\nu_k \le c_0 Mh^{\alpha}\mu_{k-1}$, then, by (9),

$$\mu_k \le \left(2c_0 Mh^{\alpha}\right)^{k-1}\mu_1 \tag{10}$$

and for sufficiently small h the errors decrease like a geometric progression.

Now we estimate the probability that the inequality (9) is valid. Applying the probabilistic assumptions of §4 to the values $\mu^{(k)}_{qj}$ and denoting the sum in (9) by $\bar{\mu}^{(k)}_{qj}$, we obtain

$$\mathrm{D}\left(\bar{\mu}^{(k)}_{qj}\right) = h^2 \sum_{q=0}^{s-1}\sum_{j=-1}^{2n-1}\left(a^{(0)}_{q_1 r_2 j_1 l_2}\right)^2 \omega_{rl}\mathrm{D}\left(\mu^{(k-1)}_{r_1 q_2 l_1 j_2}\right) \le c_1^2 M^2 h\mu_{k-1}^2 , \quad c_1 = \text{const}.$$

Since the mean of the random variables $\bar{\mu}_{qj}^{(k)}$ is assumed to be zero, it follows from Chebyschev's inequality that

$$P\left(\left|\bar{\mu}_{qj}^{(k)}\right| \le tc_1 Mh^{\frac{1}{2}}\mu_{k-1}\right) \ge 1 - t^{-2} .$$

Putting $t = \frac{c_0}{c_1} h^{\alpha-3/2}$, we find

$$P\left(\left|\bar{\mu}_{qj}^{(k)}\right| \le c_0 Mh^{\alpha-1}\mu_{k-1}\right) \ge 1 - \frac{c_1^2}{c_0^2} h^{3-2\alpha} .$$

The inequality (9) is "a practical reality".

We note that, here and in §4, the probabilistic estimate for the error decreases monotomically along with h .

§6. Some Additional Notes

a. The method introduced above is easily extended to certain multidimensional integral equations with appropriately smooth kernels. The simplest case is when the integral is defined on an m-dimensional cube. In this situation, it is sufficient to treat x and y as points of the cube in the above formulas and change in a corresponding manner the sense of the remaining notation. The extension of the present results to the multidimensional case is also possible when the region of integration can be transformed into a cube using a sufficiently smooth mapping. Examples include parrallelepipeds, m-dimensional balls, spheres, toruses, spherical layers, ellipsoids, and so on.

b. To formalise the presentation we consider the case of one independent variable. Let the resolvent of (1.1) be approximated in the manner described above. The approximate resolvent is

$$\Gamma(x, y, \lambda) \approx \frac{1}{\bar{D}^h(\lambda)} \sum_{\underline{q}=\underline{0}}^{\underline{s}-\underline{1}} \sum_{\underline{j}=-\underline{1}}^{\underline{2n}-\underline{1}} \gamma_{qj}\omega_{q_1}\left(\frac{x}{h} - j_1\right)\omega_{q_2}\left(\frac{y}{h} - j_2\right) .$$

Assuming that $D(\lambda) \ne 0$, we obtain the approximate solution of the equation

$$u(x) = f(x) + \lambda \int_0^1 \Gamma(x, y; \lambda)f(y)dy$$

$$\approx f(x) + \frac{\lambda}{\bar{D}^h(\lambda)} \sum_{\underline{q}=\underline{0}}^{\underline{s}-\underline{1}} \sum_{\underline{j}=-\underline{1}}^{\underline{2n}-\underline{1}} \gamma_{qj}\omega_{q_1}\left(\frac{x}{h} - j_1\right) \int_0^1 f(y)\omega_{q_2}\left(\frac{y}{h} - j_2\right)dy , \quad (1)$$

and we reduce the problem to calculation of integrals

$$\int_0^1 f(y)\omega_q\left(\frac{y}{h} - j\right)dy \quad (2)$$

where this time q and j are integers. If the function f is sufficiently smooth, then the integrals (2) can be calculated by the method presented in §1 of Chapter VIII. Thus, if $f(x)$ is continuous and $0 \le j \le 2n-2$, then denoting $jh = x_j$, we obtain

$$\int_0^1 f(y)\omega_q\left(\frac{y}{h} - j\right)dy = h\int_0^2 f\left(x_j+th\right)\omega_q(t)dt \approx hf\left(x_j\right)\int_0^2 \omega_q(t)dt \ . \tag{3}$$

The error of the approximation (3) has order $h\omega(f, 2h)$, where $\omega(f, \delta)$ denotes the modules of continuity of f.

c. If, by the method of §1, the Fredholm determinant $D(\lambda)$ is evaluated approximately in the form of a polynomial with respect to λ, then, applying the usual methods (for example, the Lobachevsky-Graeffe method), one can find approximations to the eigenvalues of the kernel $K(x, y)$.

d. In the preceding sections of the present chapter the approximation strategy developed for integral equations can be used to solve two-dimensional problems of potential and elasticity theory for regions bounded by sufficiently smooth curves.

§7. Equations with Weak Singularities

Let the kernel in (1.1) take the form

$$K(x, y) = \frac{A(x,y)}{r^\lambda}, \quad r = |y-x|, \quad 0 < \lambda < 1,$$
$$A \in C^{(1)}[0, 1 \quad 0, 1], \quad |A(x, y)| \le a . \tag{1}$$

We set

$$K(x, y; \varepsilon) = \begin{cases} K(x, y), & r \ge \varepsilon, \\ \dfrac{A(x,y)}{\varepsilon^\lambda}, & r < \varepsilon . \end{cases} \tag{2}$$

Obviously, the kernel (2) is continuous and is contained in $W_\infty^{(1)}$. It can be easily seen that the norm of the difference between the integral operators K and $K(\varepsilon)$ (corresponding to the kernels $K(x, y)$ and $K(x, y; \varepsilon)$), tends to zero along with ε in the spaces C and L_p. We give the proof for the space C, and at the same time derive an estimate in this norm. We have

$$\left|(Ku)(x)-\left(K(\varepsilon)u\right)(x)\right| = \left|\int_{r<\varepsilon} A(x, y)\left(r^{-\lambda}-\varepsilon^{-\lambda}\right)u(y)dy\right|$$
$$\le a\|u\|_C \int_{r<\varepsilon}\left(r^{-\lambda}-\varepsilon^{-\lambda}\right)dy = a\|u\|_C\left[\int_{r<\varepsilon} r^{-\lambda}dy-2\varepsilon^{1-\lambda}\right] .$$

Now

$$\int_{r<\varepsilon} \frac{dy}{r^\lambda} = \int_{x-\varepsilon}^{x} \frac{dy}{(x-y)^\lambda} + \int_{x}^{x+\varepsilon} \frac{dy}{(y-x)^\lambda} = \frac{2\varepsilon^{1-\lambda}}{1-\lambda} .$$

Hence

$$|(Ku)(x) - \big(K(\varepsilon)u\big)(x)| \le \frac{2\lambda a}{1-\lambda} \varepsilon^{1-\lambda} \|u\|_C$$

and, consequently,

$$\|K - K(\varepsilon)\|_C \le \frac{2\lambda a}{1-\lambda} \varepsilon^{1-\lambda} . \tag{3}$$

So, if ε is sufficiently small, the solution of the integral equation with kernel $K(x, y)$ can be replaced approximately, but with sufficient accuracy, by the solution of the integral equation with continuous kernel $K(x, y; \varepsilon)$. We noted that it is contained in $W_\infty^{(1)}$, so it can be approximated with respect to the norm in C by the kernel

$$K^h(x, y; \varepsilon) = \sum_{j,k=-1}^{2n-1} K\big((j+1)h, (k+1)h; \varepsilon\big) \omega_0\left(\frac{x}{h} - j\right) \omega_0\left(\frac{y}{n} - j\right) \tag{4}$$

where $h = 1/2n$ with n a positive integer and

$$\omega_0(t) = \begin{cases} t , & 0 \le t \le 1 , \\ 2-t , & 1 < t \le 2 , \\ 0 , & t \notin [0, 2] . \end{cases}$$

The norm of the difference $K - K(h, \varepsilon)$, where K and $K(h, \varepsilon)$ are integral operators with kernels $K(x, y)$ and $K^h(x, y; \varepsilon)$, respectively, is arbitrarily small for sufficiently small values of ε and h , and the resolvent of the kernel $K^h(x, y; \varepsilon)$ can be constructed by the method of §1 or, for small values of the Fredholm parameter, by the method of §4.

§8. Integral Equations of Heat Conduction

Let Ω be a bounded region on the plane (x, y) and Γ its boundary. We seek the solution of the heat conduction equation

$$\frac{\partial u}{\partial t} = \frac{\partial^2 u}{\partial x^2} + \frac{\partial^2 u}{\partial y^2}$$

in the region Ω for time $t > 0$ for the initial condition $u|_{t=0} = 0$ and the boundary condition $u|_\Gamma = \varphi(s, t)$, where s is a parameter defining the position of a point on Γ . We assume that $0 \le s \le 1$. We seek the solution in the form of the heat potential of a double layer (see, for example, Müntz [1]):

$$u(x, y, t) = \frac{1}{2\pi}\int_0^t \left\{\int_0^1 \frac{\mu(\sigma,\tau)}{t-\tau}\frac{\partial}{\partial\nu}\exp\left(-\frac{r^2}{4(t-\tau)}\right)d\sigma\right\}d\tau .$$

Here σ is the value of the parameter s, corresponding to the point of integration on the contour Γ, ν is the outward normal to Γ at the point σ, $\mu(\sigma, \tau)$ is the unknown density of the potential, r is the distance from a point $(x, y) \in \Omega$ to the point on Γ, corresponding to the value σ. The density satisfies the integral equation

$$\mu(s, t) - \frac{1}{4\pi}\int_0^t \left\{\int_0^1 \frac{\mu(\sigma,\tau)}{(t-\tau)^2}\exp\left(-\frac{r^2}{4(t-\tau)}\right)r\cos(r, v)d\sigma\right\}d\tau = -\varphi(s, t) , \tag{1}$$

where r is the distance between points s and σ on Γ and at the same time the vector from s to σ. The equation (1) has a unique solution which can be found by iteration. The convergence of the iteration is studied in Müntz [1] for a sufficiently smooth contour, and in Mikhlin [20] for a convex contour.

We assume that the curve Γ has continuously differentiable curvature, then $\cos(r, v) = r\cdot A(s, \sigma)$, where A is a continuously differentiable function.

We replace the denominator $(t-\tau)^2$ in (1) by $(t-\tau)^2 + \varepsilon^2$, where ε is an arbitrary positive number. Denoting the unknown of the new equation by μ_ε, we obtain

$$\mu_\varepsilon(s, t) - \frac{1}{4\pi}\int_0^t \left\{\int_0^1 \frac{\mu_\varepsilon(\sigma,\tau)}{(t-\tau)^2+\varepsilon^2}\exp\left(-\frac{r^2}{4(t-\tau)}\right)r^2A(s, \sigma)d\sigma\right\}d\tau = -\varphi(s, t) . \tag{2}$$

We prove that the norm of the difference between the operators (1) and (2) tends to zero along with ε with respect to the metrics of $C([0, \infty) \times [0, 1])$, which consists of continuous and bounded functions on the strip $0 \le t < \infty$, $0 \le s \le 1$. Let $\mu(s, t)$ be such a function. We set $A_0 = \max|A(s, \sigma)|$, so that

$$\left|\frac{1}{4\pi}\int_0^t \left\{\int_0^1 \mu(\sigma, \tau)\left[\frac{1}{(t-\tau)^2} - \frac{1}{(t-\tau)^2+\varepsilon^2}\right]r^2A(s, \sigma)\exp\left(-\frac{r^2}{4(t-\tau)}\right)d\sigma\right\}d\tau\right|$$

$$\le A_0\|\mu\|_C\varepsilon^2\int_0^1\left\{\int_0^t \frac{r^2}{\tau^2(\tau^2+\varepsilon^2)}\exp\left(-\frac{r^2}{4\tau}\right)d\tau\right\}d\sigma .$$

Thus, the norm of the difference we are interested in is dominated by

$$M = A_0\varepsilon^2\int_0^1\left\{\int_0^\infty \frac{r^2}{\tau^2(\tau^2+\varepsilon^2)}\exp\left(-\frac{r^2}{4\tau}\right)d\tau\right\}d\sigma .$$

Introducing the change of variables defined by $r^2/4\tau = \xi$, we obtain

$$M = 64A_0\varepsilon^2\int_0^1\left\{\int_0^\infty \frac{\xi^2e^{-\xi}d\xi}{r^2+16\varepsilon^2\xi^2}\right\}d\sigma = 64A_0\varepsilon^2\int_0^\infty \xi^2e^{-\xi}\left\{\int_0^1 \frac{d\sigma}{r^2+16\varepsilon^2\xi^2}\right\}d\xi . \tag{3}$$

Let the beginning of arcs on Γ coincide with s , so that $r = 0$ when $\sigma = 0$ and $\sigma = 1$. It can be easily verified that $r = \sigma(1-\sigma)f(\sigma)$, where $f(\sigma)$ is bounded and its upper and lower bounds are positive numbers. Let $f(\sigma) \geq C_0^{-1}$ = const.; then

$$\int_0^1 \frac{d\sigma}{r^2+16\varepsilon^2\xi^2} \leq C_0^2 \int_0^1 \frac{d\sigma}{\sigma^2(1-\sigma)^2+16C_0^2\varepsilon^2\xi^2} .$$

We interpret the r.h.s. as the sum of two integrals over the intervals $(0, \frac{1}{2})$ and $(\frac{1}{2}, 1)$. For the first integral, we have $1 - \sigma \geq \frac{1}{2}$, and therefore it is dominated by

$$4 \int_0^{\frac{1}{2}} \frac{d\sigma}{\sigma^2+64C_0^2\varepsilon^2\xi^2} = \frac{1}{2C_0\varepsilon\xi} \operatorname{arctg} \frac{\sigma}{8C_0\varepsilon\xi}\bigg|_0^{\frac{1}{2}} \leq \frac{\pi}{4C_0\varepsilon\xi} .$$

The same estimate is also valid for the second integral. Hence

$$\int_0^1 \frac{d\sigma}{r^2+16\varepsilon^2\xi^2} < \frac{C_0\pi}{2\varepsilon\xi} .$$

By (3),

$$M < 32A_0\pi C_0\varepsilon \int_0^\infty \xi e^{-\xi} d\xi = 32A_0\pi C_0\varepsilon ,$$

and the result follows. So, (2) has for sufficiently small ε the unique solution $\mu_\varepsilon(s, t)$ and the difference $\mu_\varepsilon(s, t) - \mu(s, t)$ tends uniformally to zero along with ε .

Below, we assume that Γ not only has continuous curvature, but also is convex. From the results of Mikhlin [21] it follows that the norm of the integral operator in (1) is less than unity with respect to metric of $C\big([0, \infty) \times [0, 1]\big)$. Then the norm of the integral operator in (2) is also less than unity for sufficiently small ε .

Take the Laplace transform of (2). Let p denote a complex variable, $p = \xi + i\eta$ with $\xi > 0$. We multiply both sides of (2) by e^{-pt} and integrate with respect to $t \in (0, \infty)$. Equation (2) becomes

$$\tilde{\mu}_\varepsilon(s, p) - \int_0^1 K(s, \sigma; p, \varepsilon)\tilde{\mu}_\varepsilon(\sigma, p)d\sigma = -\tilde{\varphi}(s, p) \tag{4}$$

where $\tilde{\mu}_\varepsilon$ and $\tilde{\varphi}$ denote the Laplace transforms for μ_ε and φ , respectively, and

$$K(s, \sigma \ \ p, \varepsilon) = \frac{r^2}{4\pi} A(s, \sigma) \int_0^\infty \frac{1}{t^2+\varepsilon^2} \exp\left(-pt - \frac{r^2}{4t}\right) dt . \tag{5}$$

We prove that (4) can be solved by iteration. It is sufficient to prove that

$$\int_0^1 |K(s,\ \sigma\ \ p,\ \varepsilon)|d\sigma \leq q = \text{const} < 1 \ . \tag{6}$$

We note that $A(s, \sigma) = d\alpha/d\sigma$, where $d\alpha$ is the angle subtended at the point s by the element $d\sigma$ on the contour. Hence,

$$\int_0^1 |K(s,\ \sigma,\ p,\ \varepsilon)|d\sigma = \frac{1}{4\pi}\int_\Gamma \left|\int_0^\infty \frac{r^2}{t^2+\varepsilon^2} \exp\left(-pt - \frac{r^2}{4t}\right)dt\right| d\alpha$$

$$\leq \frac{1}{4\pi}\int_\Gamma \left\{\int_0^\infty \frac{r^2}{t^2+\varepsilon^2} \exp\left(-\xi t - \frac{r^2}{4t}\right)dt\right\}d\alpha \ . \tag{7}$$

We divide the inner integral into the sum of two integrals over the intervals $(0,\ \delta)$ and $(\delta,\ \infty)$, where δ is an arbitrary positive number. We have

$$\int_\delta^\infty \frac{r^2}{t^2+\varepsilon^2} \exp\left(-\xi t - \frac{r^2}{4t}\right)dt < e^{-\xi\delta}\int_\delta^\infty \frac{r^2}{t^2} \exp\left(-\frac{r^2}{4t}\right)dt = 4e^{-\xi\delta}\left(1-e^{-r^2/4\delta}\right) \ .$$

In addition,

$$\int_0^\delta \frac{r^2}{t^2+\varepsilon^2} \exp\left(-\xi t - \frac{r^2}{4t}\right)dt < 4e^{-r^2/4\delta} \ .$$

As a result, the inner integral can be estimated as

$$\int_0^\infty \frac{r^2}{t^2+\varepsilon^2} \exp\left(-\xi t - \frac{r^2}{4t}\right)dt < 4\left[e^{-\xi\delta}+e^{-r^2/4\delta}-e^{-\xi\delta-r^2/4\delta}\right] \ . \tag{8}$$

We put $\delta = r/2\gamma$ and $\gamma = \sqrt{\xi}$. Then the r.h.s. in (8) equals $4\left(2e^{-\gamma r/2}-e^{-\gamma r}\right)$. Hence

$$\int_0^1 |K(s,\ \sigma\ \ p,\ \varepsilon)|d\sigma < \frac{1}{\pi}\int_\Gamma \left(2e^{-\gamma r/2}-e^{-\gamma r}\right)d\alpha \overset{\text{def}}{=\!=} \kappa(s) \ . \tag{9}$$

The function under the integral sign on the r.h.s. of (9) equals one when $r = 0$, and is contained in the interval $(0,\ 1)$ when $r > 0$. Hence, it follows that, for any $s \in \Gamma$, the function $\kappa(s)$ is less than one. The function $\kappa(s)$ is continuous, which follows from

$$\kappa(s) = \frac{1}{\pi}\int_0^1 A(s,\ \sigma)\left(2e^{-\gamma r/2}-e^{-\gamma r}\right)d\sigma$$

and from the continuity of the function $A(s,\ \sigma)$. Hence,

$$q \overset{\text{def}}{=\!=} \max \kappa(s) < 1 \ . \tag{10}$$

The inquality (6) is thereby verified. We note that q depends neither on ε nor on η and that (10) is valid for any $\xi > 0$. It can be easily seen that the kernel (5) is bounded independently of ξ. If $\xi > 0$ and Γ is sufficiently smooth, then the continious differentiability of (5) with respect to s and σ can be guaranteed up to an appropriate number of times.

It follows from (6) that the smallest characteristic value of the kernel (5) is greater than $q^{-1} > 1$. Consequently, the resolvent of (4) can be expressed as a series in terms of iterated kernels and that this series converges uniformly with respect to s, σ and ε . The resolvent can also be expressed as the quotient of Fredholm series, which also converge uniformly with respect to the mentioned variables, where the denominator of the resolvent has a larger absolute magnitude than some positive constant independent of ε and σ . It implies that one can apply the methods of §§1 and 5 to (4). As an example, we consider a rather rough but simple approximation. A more accurate approximation can be constructed using a primitive system of non-zero degree. Let the primitive function $\omega_0(t)$ be defined by (6.5) of Chapter II. We choose positive integer n and put $h = 1/2n$. The kernel (5) can be approximated as

$$K(s, \sigma; p, \varepsilon) \approx K^h(s, \sigma; p, \varepsilon) = \sum_{j,k=-1}^{2n-1} a_{jk}^{(0)}(p, \varepsilon)\omega\left(\frac{x}{h} - j\right)\omega_0\left(\frac{\sigma}{h} - k\right) , \tag{11}$$

where

$$\begin{aligned} a_{jk}^{(0)}(p, \varepsilon) &= K\big((j+1)h, (k+1)h; p, \varepsilon\big) \\ &= \frac{1}{4\pi} A\big((j+1)h, (k+1)h\big)r_{jk}^2 \int_0^\infty \frac{1}{t^2+\varepsilon^2} \exp\left(-pt - \frac{r_{jk}^2}{4t}\right)dt , \end{aligned} \tag{12}$$

and r_{jk} is the distance between the points $s = (j+1)h$ and $\sigma = (k+1)h$. The error in the approximation (11) is of order $O(h^2)$.

The difference $j - k$ is contained in the interval $[-2n, 2n]$ and $a_{jj} = 0$. The application of (11) involves the evaluation of (12), where N is the number of values of η for which we wish to solve (4). We note that these integrals converge sufficiently rapidly and their evaluation is not very labourous.

Let $\Gamma^h(s, \sigma; p, \varepsilon)$ denote the resolvent of (11), when the Fredholm parameter equals one. This resolvent can be expressed as the quotient of the Fredholm series

$$\Gamma^h(s, \sigma; p, \varepsilon) = \frac{D^h(s,\sigma;p,\varepsilon)}{D^h(p,\varepsilon)} ,$$

where

$$D^h(p, \varepsilon) = \sum_{l=0}^{\infty} \frac{(-1)^l}{l!} C_l^h(p, \varepsilon) ,$$

$$D^h(s, \sigma; p, \varepsilon) = \sum_{l=0}^{\infty} \frac{(-1)^l}{l!} B_l^h(s, \sigma; p, \varepsilon) ,$$

and C_l^h, B_l^h satisfy the recurrence relations

$$C_0^h(p, \varepsilon) = 1 ,$$

$$C_l^h(p, \varepsilon) = \int_0^1 B_{l-1}^h(s, s; p, \varepsilon)ds , \quad l > 0 ,$$

$$B_0^h(s, \sigma; p, \varepsilon) = K^h(s, \sigma; p, \varepsilon) , \tag{13}$$

$$B_l^h(s, \sigma; p, \varepsilon) = C_l^h(p, \varepsilon)K^h(s, \sigma; p, \varepsilon) - l\int_0^1 K^h(s, \lambda; p, \varepsilon)B_{l-1}^h(\lambda, \sigma; p, \varepsilon)d\lambda , \quad l > 0 .$$

It follows from (13) that

$$B_l^h(s, \sigma; p, \varepsilon) = \sum_{j,k=-1}^{2n-1} a_{jk}^{(l)}(p, \varepsilon)\omega_0\left(\frac{s}{h} - j\right)\omega_0\left(\frac{\sigma}{h} - k\right) , \tag{14}$$

$$C_l^h(p, \varepsilon) = h\sum_{j,k=-1}^{2n-1} a_{jk}^{(l-1)}(p, \varepsilon)w_{jk} ,$$

where

$$hw_{jk} = \int_0^1 \omega_0\left(\frac{s}{h} - j\right)\omega_0\left(\frac{s}{h} - k\right)ds , \tag{15}$$

and the coefficients $a_{jk}^{(l)}$ can be evaluated by the recurrence formula

$$a_{jk}^{(l)} = C_l^h a_{jk}^{(0)} - lh\sum_{\mu,\nu=-1}^{2n-1} a_{j\nu}^{(0)}a_{\mu k}^{(l-1)}w_{\mu\nu} . \tag{16}$$

The integrals (15) can be readily evaluated as

$$w_{jk} = \begin{cases} 0 , & |j-k| > 1 , \\ 1/3 , & j = k = -1, 2n-1 , \\ 2/3 , & j = k = 0, 1, \ldots, 2n-2 , \\ 1/6 , & j = k \pm 1 . \end{cases}$$

When the resolvent is constructed, the approximate solution of equation (4) can be easily determined. We then apply the inverse Laplace transform and find the required density $\mu_\varepsilon(s, t)$; the results of Riabov [1], as well as the results of works cited there, can be useful.

The method and the results of the present section can be extended without any essential changes to the situations where the boundary conditions of the second or third boundary value problem are considered on Γ . It is only necessary to replace the potential of a double layer by the analogous potential for a simple layer.

We examined the equation for heat conduction in two variables. The extension to multidimensions involves no difficulties except for one point: if the number of variables $m > 2$, then the method of §1 demands (as it was noted in §6) that the boundary Γ of Ω be maped sufficiently smoothly onto a cube.

REFERENCES

AUBIN, J.-P.

[1] Évaluation des erreurs de troncature des approximations des espaces de Sobolev. *J. Math. Anal. Appl.*, 21, 1968, 356-368. (MR36#6923)

BABITCH, V.M. (Бабич, В.М.)

[1] On extension of functions. *Usp. Mat. Nauk* 8:2 (54), 1953, 111-113 (Russian). (MR15,110)

BABUŠKA, I., E, VITÁSEK, M. PRÁGER

[1] *Numerical Processes in Differential Equations* (translated from Czech). Interscience Publishers (John Wiley & Sons), London-New York-Sydney, 1966, x + 351 pp. (MR36#6150)

BARI, N.K. (Бари, Н.К.)

[1] Generalization of inequalities of S.N. Bernschtein and A.A. Markov. *Dokl. Akad. Nauk SSSR* (N.S.) 90, 1953, 701-702 (Russian). (MR15,215)

[2] Generalization of inequalities of S.N. Bernschtein and A.A. Markov. *Izv. Akad. Nauk SSSR, Ser. Mat.* 18, 1954, 159-176 (Russian). (MR15,788)

BIRMAN, M.S., M.Z. SOLOMJAK (Бирман, М.Ш., М.З. Соломяк, М.З.)

[1] On the principal term of a spectral asymptotic for "nonsmooth" elliptic problems. *Funk. Anal. Prilozen* 4:4, 1970, 1-13 (Russian). (MR43#3857)

CALDERON, A.P.

[1] Lebesgue spaces of differentiable functions and distributions. Proc. Sympos. Pure Math., Vol. IV, pp. 33-49. Amer. Math. Soc., Providence, Rhode Island, 1961. (MR26#603)

CALDERON, A.P., A. ZYGMUND

[1] On singular integrals. *Amer. J. Math.* 78, 1956, 289-309. (MR18,894)

CIMMINO, G.

[1] Nuovo tipo di condizione al contorno e nuovo metodo di trattazione per il problema generalizzato di Dirichlet. *Rend. Circ. Mat. Palermo*, 61A, 1937, 177-221.

COURANT, R.

[1] Variational methods for the solution of problems of equilibrium and vibrations. *Bull. Amer. Math. Soc.*, **49**, 1943, 1-23. (MR4,200)

DEM'JANOVIC, J.K. (Демьянович, Ю.К.)

[1] The stability of the net method for elliptic problems. *Dokl. Akad. Nauk SSSR*, **164**, 1965, 20-23 (Russian). (MR32#6689)

DEM'JANOVIC, J.K., S.G. MICHLIN (Демьянович, Ю.К., С.Г. Михлин)

[1] On finite-difference approximation of functions from Sobolev classes. Zapiski nauch. seminar. *Akad. Nauk SSSR*, **35**, 1973, 6-11.

FICHTENHOLZ, G.M.

[1] *Differential- und Integralrechnung*, Bd. I. VEB Deutscher Verlag der Wissenschaften, Berlin 1964.

FORSYTHE, G.E., W.R. WASOW

[1] *Finite-Difference Methods for Partial Differential Equations*. John Wiley & Sons, New York-London, 1960, x + 444pp. (MR23#B3156)

FRANK, L.S. (Франк, Л.С.)

[1] Difference convolution operators. *Dokl. Akad. Nauk SSSR*, **181**, 1968, 286-289.

GAVURIN, M.K. (Гавурин, М.К.)

[1] *Lectures on Numerical Methods*. Nauka, 1971.

GODUNOV, S.K., V.W. RJABENKIY

[1] *Theory of Difference Schemes. An Introduction*. Interscience Publishers (John Wiley & Sons), New York, 1964, xii + 289 pp. (Russian). (MR31#5346)

GOËL, J.-J.

[1] Construction of basis functions for numerical utilisation of Ritz's Method. *Numer. Math.*, **12**, 1968, 435-447. (MR41#1236)

GOLOVKIN, K.K. (Головкин, К.К.)

[1] On approximation of functions in arbitrary norms. *Trudi Mat. Inst. Steklov*, **70**, 1964, 26-37 (Russian). (MR28#5281)

GUSMAN, Ju.A., L.A. OGANESJAN (Гусман, Ю.А., Оганесян, Л.А.)

[1] Estimates for the convergence of finite-difference schemes for degenerate elliptic equations. *Z. Vycisl. Mat. i Mat. Fiz.*, **5**, 1965, 351-357 (Russian) (MR33#6851)

IL'IN, V.P. (Ильин, В.П.)

[1] Properties of some classes of differentiable functions of many variables defined on a given n-dimensional region. *Trudy Matem. in-ta im. VA Steklov ANSSSR*, **66**, 1962, 227-236.

KRASNOSELSKIJ, M.A., u.a.

[1] *Näherungsverfahren zur Lösung von Operatorgleichungen.* Akademie-Verlag, Berlin, 1973.

LOVITT, W.V.

[1] *Linear integral equations.* McGraw-Hill, New York, 1924.

MIKHLIN, S.G. (Михлин, С.Г.)

[1] *The Problem of the Minimum of a Quadratic Functional* (translated from Russian). Holden-Day, Inc., San Francisco, 1965, ix + 155 pp. (MR30#1427)

[2] Degenerate elliptic equations. *Vestnik Leningrad Univ.*, **9**:8, 1954, 19-48 (Russian). (MR17,493)

[3] *Variational Methods in Mathematical Physics* (translated from Russian). A Pergomon Press Book, The Macmillan Co., New York, 1964, xxxiii + 582 pp. (MR30#2712)

[4] *Multidimensional Singular Integrals and Integral Equations* (translated from Russian). Pergamon Press, Oxford-New York-Paris, 1965, xi + 255 pp. (MR32#2866)

[5] *The Numerical Performance of Variational Methods* (translated from Russian). Wolters-Noordhoff Publishing, Groningen, 1971, xxiii + 373 pp. (MR43#4236)

[6] *Mathematical Physics. An Advanced Course* (translated from Russian). North Holland Publishing Co., Amsterdam-London, 1970, xv + 561 pp. (MR44#3538)

[7] The variational-difference method for one-dimensional boundary value problems. *Dokl. Akad. Nauk SSSR*, **198**, 1971, 39-41 (Russian). (MR44#2354)

[8] The coordinate systems of the variational-difference method. *Dokl. Akad. Nauk SSSR*, **200**, 1971, 526-529 (Russian). (MR44#4580)

[9] The variational-difference method for multi-dimensional boundary value problems. *Zap. Naucn. Sem. Leningrad Otdel. Mat. Inst. Steklov (LOMI)*, **23**, 1971, 99-114 (Russian). (MR46#1102)

[10] Some properties of variational-difference schemes for one dimensional problems. Sbornik "Mechan. sploshnoi sredi i rodstv. problem. anal.". Nauka, 1972, 311-321.

[11] Certain questions of approximation in the variational difference method. *Dokl. Akad. Nauk SSSR*, **209**, 1973, 299-301 (Russian). (MR48#1489)

[12] The finite-difference approximation of the solutions of degenerate one-dimensional differential equations of order 2 . *Vestnik. Lening. Univ.* No 1, *Mat. Meh. Astronom. Vyp.*, **1**, 1973, 52-67 (Russian). (MR48#5391)

[13] Approximation on a radial-annular net. *Zap. Naucn. Sem. Leningrad Otdel. Mat. Inst. Steklov (LOMI)*, **35**, 1973, 95-102 (Russian). (MR48#12778)

[14] On the smallest number of original functions of the variational-difference method. *Zap. Naucn. Sem. Leningrad Otdel. Mat. Inst. Steklov (LOMI)*, **35**, 1973, 103-105.

[15] The condition number of the variational-grid-matrix. *Vestnic. Lening. Univ.* No. 13, *Mat. Meh. Astronom. Vyp.*, **3**, 1973, 162-164 (Russian). (MR48#7624)

[16] On one class of high accuracy variational-difference schemes. "Issledovania po teorii sooruzenii", XXI, Gosstroiizdat, 1974, 9-18.

[17] A certain class of coordinate functions of the variational-difference method. *Dokl. Akad. Nauk SSSR*, **211**, 1973, 1057-1059 (Russian). (MR48#10145)

[18] On the variational-difference method. *Vestnic. Lening. Univ.*, No. 1, 1974, 40-47.

[19] On one method of approximate solution of integral equations. *Vestnic. Lening. Univ.*, No. 13, 1974, 26-33.

[20] On the approximate solution of the heat-conduction integral equations. Rendiconti di Matematica, 1975.

[21] Applications of integral equations. CTTI, 1947.

MÜNTZ, H.M. (Мюнтц, Г.М.)

[1] *Integral Equations*. Gostechisdat, 1934.

NAIMARK, M.A.

[1] *Linear Differential Operators* (translated from Russian). Frederick Ungar Publishing Co., New York, 1967 and 1968, xxiii + 144 pp. (Part I), xv + 352 pp. (Part II). (MR35#6885, MR41#7485)

NIKOLSKY, S.M. (Никольский, С.М.)

[1] On the solution of the polyharmonic equation by a variational method. *Dokl. Akad. Nauk SSSR* , **88**, 1953, 409-411 (Russian). (MR15,425)

OGANESJAN, L.A. (Оганесян, Л.А.)

[1] Variational-difference schema for the Dirichlet problem on the regular net. *Z. Vytisl. Mat. i Mat. Fiz.*, **11**, 1971, 1595-1603 (Russian). (MR45#4666)

RIABOV, V.M. (Рябов, В.М.)

[1] On the numerical inversion of the Laplace transform. *Vestnic Lening. Univ.*, No. 7, 1974, 68-75.

RUKHOVETZ, L.A. (Руховец, Л.А.)

[1] On the composition of variational-difference schemes for elliptic equations. *Z. vichisl. mat. i mat. phys.* **12**:3, 1972, 781-785.

SHILOV, G.E. (Шилов, Г.Е.)

[1] *Mathematical Analysis. A Special Course* (translated from Russian). Pergamon Press, Oxford-New York-Paris, 1965, xii + 481 pp. (MR32#2519)

SMIRNOV, V.I.

[1] Lehrgang der hoheren Mathematik, Bd. IV. VEB Deutscher Verlag der Wissenschaften, Berlin, 1966.

SOBOLEV, S.L. (Соболев, С.Л.)

[1] *Applications of Functional Analyses in Mathematical Physics* (translated from Russian). Translations of Mathematical Monographs, Vol. 7, American Mathematical Society, Providence, Rhode Island, 1963, vii + 239 pp. (MR29#2624)

[2] *Introduction to the Theory of Cubature Formulas*. Nauka, 1974.

STRANG, G.

[1] Approximation in the finite element method. *Numer. Math.*, **19**, 1972, 81-98.

STRANG, G., G. FIX

[1] Fourier analysis of the finite element in Ritz-Galerkin theory. *Stud. Appl. Math.*, **48**, 1969, 265-273. (MR41#2944)

[2] A Fourier analysis of the finite element method. A preprint of: *The Finite Element Method*.

TRICOMI, G.F.

[1] *Integral Equations*. Interscience Publishers, New York, London, 1957.

VAINIKKO, G.M. (Вайникко, Г.М.)

[1] Certain estimates of the error in the Bubnov-Galerkin method. II. Estimates of the nth approximation. *Tertu Riikl Ul Toimetised*, No. 150, 1964, 202-215 (Russian). (MR33#68656)

[2] Estimates of the error of the Bubnov-Galerkin method in an eigenvalue problem. *Z. Vycsil Mat. i Mat. Fiz.* **5**, 1965, 587-607 (Russian). (MR34#5271)

VARGA, R.S.

[1] *Functional Analysis and Approximation Theory in Numerical Analysis* . Soc. for Industr. and Appl. Mathem., Philadelphia, 1971, v + 76 pp. (MR46#9602)

WHITTAKER, E.T., G. ROBINSON

[1] *The Calculus of Observations: A Treatise on Numerical Mathematics*, Fourth Edition. Dover Publications Inc., New York, 1967, xiv + 397 pp. (MR38#1087)

WILKINSON, J.H.

[1] *The Algebraic Eigenvalue Problem*. Clarendon Press, Oxford, 1965, xviii + 662 pp. (MR32#1894)

INDEX